Leadership in American Academic Geography

Leadership in American Academic Geography

The Twentieth Century

Michael S. DeVivo

LEXINGTON BOOKS
Lanham • Boulder • New York • London

Published by Lexington Books
An imprint of The Rowman & Littlefield Publishing Group, Inc.
4501 Forbes Boulevard, Suite 200, Lanham, Maryland 20706
www.rowman.com

16 Carlisle Street, London W1D 3BT, United Kingdom

British Library Cataloguing in Publication Information Available

Library of Congress Cataloging-in-Publication Data

DeVivo, Michael S.
Leadership in American academic geography : the twentieth century / Michael S. DeVivo.
pages cm
Includes bibliographical references and index.
ISBN 978-0-7391-9912-1 (cloth : alk. paper) -- ISBN 978-0-7391-9913-8 (electronic)
1. Geography--Study and teaching (Higher)--United States--History--20th century. 2. Geographers--United States--Biography. I. Title.
G76.5.U5D48 2014
910.71'173--dc23
2014033126

∞™ The paper used in this publication meets the minimum requirements of American National Standard for Information Sciences Permanence of Paper for Printed Library Materials, ANSI/NISO Z39.48-1992.

Printed in the United States of America

To my parents,

Albert and Ruth DeVivo

Contents

Preface ix

Acknowledgments xiii

1 Introduction 1

2 The Ascent of American Academic Geography 13

3 Transformational Leadership at Chicago: The Post-Salisbury Years 21

4 Paradoxical Leadership at Wisconsin 25

5 Lackluster Leadership at Michigan 35

6 From Tyranny to Transformational Leadership at Minnesota 41

7 The Struggle for Distinction at Ohio State 49

8 The Transformational Leadership of E. Willard Miller and Penn State Geography 55

9 George Cressey and Preston James at Syracuse 61

10 The Transactional Leadership of Wallace Atwood and the Emergence of Geography at Clark 67

11 Duplicity and Deception at Johns Hopkins 73

12 Laissez Faire Leadership at Harvard and Geography's Demise 83

13 G. Donald Hudson's Transformational Leadership at Northwestern 93

14 G. Donald Hudson's Transformational Leadership at Washington 99

15 Iowa's Rise to Prominence 105

16 Transformational Leadership at UCLA 111

17 The Legacy of Carl Sauer: Transformational Leadership at Berkeley 117

18 Leaders in a Paradigm of Eclectic Pluralism 127

19 Simonett and the Santa Barbarians 149

20 The Transformational Leadership Imperative 159

21 Epilogue 167

Appendix A: Leadership in Academic Departments: A Review 183
Appendix B: Sources on the History of Geography 185
References 193
Index 207
About the Author 211

Preface

> Pleasantly zoned on the beach, Golledge was extracted from reverie by someone kicking sand on him and barking the traditional Australian greeting of "Ow are ya, mate?" Turning his head to the left, the sunbather spotted a pair of white freckled legs, pants rolled up to the knees. Looking upwards, Golledge recognized the smiling face of David Simonett, who immediately asked, "Have you made up your mind yet? When are you coming to Santa Barbara?" Without waiting for an answer, Simonett crouched down and launched into the plans he had for the fledgling Geography Department at the University of California Santa Barbara, what Golledge was going to teach, specifics about salary and health benefits. He drew in the sand to illustrate and spoke as if it were a foregone conclusion Golledge would come (Baumgart 2004, p. 1).[1]

At a national conference in 2006, I related this story to a full house of academic geographers while delivering a paper, using my best Aussie brogue where apropos, when in the back of the room, a heavy-set gentleman shouted, "That's *exactly* how it happened!" It was Reg Golledge that made the remark. I had never met him before, and after my presentation on the emergence and decline of American academic departments of geography, he was among the several individuals that approached me. As a former President of the Association of American Geographers, he stated that my work needed to be published; it was imperative for the community of academic geographers to read my assessment of leadership throughout the discipline's history. Susan Hardwick, a former President of the National Council for Geographic Education agreed, and Mary Lynne Byrd, the Executive Director of the American Geographical Society also concurred, and asked when the work would be published.

To say that I was pleasantly surprised by the response my work generated is an understatement. Fortuitous circumstances placed us all in the same meeting room in an out-of-the-way location on the eastern side of the Sierras, and the favorable comments, compelled me to ponder this matter further. The field of the *history of geographic thought* had always intrigued me. I had over the course of more than a couple of decades conducted much study and engaged in dialogue with several of the scholars that had developed mastery of the field, and I had spent considerable time talking with geographers that played parts in the evolution of the discipline during the twentieth century. Moreover, the matter of scholarship devoted to the discipline's leadership had largely been ignored, and the need to address this inadequacy was dire. The die was cast. The lot fell to me. And what began as a brief research article evolved into a book.

Unfortunately, Golledge, who is partly responsible for inspiring this effort because of his adamant stance that it be completed, passed away before it was done. The Chancellor at UC-Santa Barbara commented, "Reg was a true visionary who always saw great possibilities and provided the insight and leadership to take advantage of them and help others." [2] Indeed, he was a "powerhouse" in academic geography and a transformational leader; our discipline sorely needs more like him.

Although he did not know this work was in progress, Wes Dow, the architect of *Geographers on Film*, "one of the richest, discipline-based oral history resources in the academic world," [3] also served as a catalyst, for not only have his numerous interviews lent keen insight, long before this research was formalized he and I engaged in frequent extended talks about the personalities that shaped the geographic discipline in America. He was gifted in his knowledge of the field, and capable of talking with geographers of all genres, genders, and eras, creating a record of the discipline's course throughout the twentieth century, which was marked by rich substance and impartiality. Regrettably, Wes passed away in 2011.

Although Reg and Wes occupied two different paradigms, both were highly respected in the discipline, as they were marked by integrity in scholarship and character, as well as an evenhandedness that is an essentiality, but all too often absent, in the discipline. It has been my intention in carrying out this research to throw preconceived notions aside and to emulate their ways. Thus, my assessment of leadership in the discipline was done with a critical eye and frankness. At some turns in the course of this research I've been taken by surprise, and I suspect others will be surprised by some of these findings as well.

Regardless, the conclusions drawn are derived from integration and analysis of numerous dialogues with many geographers, a number of whom have passed on, written accounts of many individuals' experiences and related biographical information, and filmed interviews conducted by Dow. My goal has been to present a fair and balanced account of the history of American academic geography in reference to its leaders in the graduate programs that most profoundly influenced its trajectory, in hopes that it contributes to the historical record of our discipline and while also offering some guidance to department chairs today and in the future. Anticipating disagreements, my hope is that this work sparks discussion and debate. Any errors are my own.

I am not connected to any of the graduate departments or personalities examined, whether as an alumnus, a professor, or through any other means; nor do I seek political office in the discipline's professional societies. Thus, I have no reason to offer either criticism or praise unless it is warranted. I simply have sought to create a truthful account of leadership in the programs that played the most influential roles in the shaping of the discipline.

NOTES

1. This incident is also discussed in Chapter 19: *Simonett and the Santa Barbarians.*
2. Henry T. Yang, University of California at Santa Barbara Chancellor.
3. Douglas Richardson, Association of American Geographers Executive Director.

Acknowledgments

Looking backward, this book, which was not really intended until Reg Golledge approached me in 2006, has actually been a work in progress for nearly thirty years; it was then that I was introduced to the history of geographic thought by Geoffrey Martin. It was through him, arguably the leading figure in the history of geographic thought in the English-speaking world, that my interest in this field was cultivated, and it was also through him that, as an aspiring geographer, I met luminaries such as Preston James, Richard Hartshorne, and Harm De Blij. Frequently, we engaged in dialogue about the shaping of the discipline, particularly in reference to individual players cast in various roles throughout the course of its history, and it was during the course of many conversations that I learned details about the leadership behaviors that characterized various individuals.

Other students of the history of our craft also must be recognized. The late Wes Dow, our discipline's oral historian *par excellence,* often conversed with me; there has been no one that has contributed more to the history of geography in North America and his influence has been significant. Martin Kenzer, now retired, also contributed substantially to scholarship here and sporadically engaged in dialogue years ago, as did Kent Mathewson, who continues to add to the large volume of work he has already accomplished. Sandy Simonett-Henderson, the widow of David Simonett, has shared keen insight, which is much appreciated, as well.

Academic geographers, luminaries themselves, that over the years were willing to spend time talking with me about the shaping of our discipline and the people that marked the past include a number of deceased members of our profession. Many contributed to the conversation; however, those that are recalled best include: Martin Glassner, Fred Kniffen, James Parsons, Robert West, Chauncy Harris, John Borchert, Cotton Mather, Everett Smith, Sam Hilliard, E.W. Miller, Reginald Golledge, and Salvatore Natoli.

There are others that, despite having witnessed more than a half-century of academic geography each, continue to relentlessly toil in the practice of the geographer's craft. My talks with Carl Johannessen, William Denevan, and Robert Richardson have generated thoughts and ideas that are written in this work. I have also conversed with: Waldo Tobler, William Koelsch, John Adams, John Boland, Saul Cohen, James Newman, Allen Scott, Ron Abler, Yi-Fu Tuan, Duane Marble, David

Ward, and Brian Berry. Each has given me valuable insight, and to say that I have been privileged to engage in dialogue with all is an understatement. I have indeed been honored, and for that I am truly grateful.

It was thirty-one years ago that I decided to become a geographer, following time in the armed forces as well as service as a U.S. Merchant Marine Officer, and it was then that I met a young assistant professor who became over the course of the last three decades, a mentor, colleague, and very dear friend. Since the recent passing of Harm de Blij, there may not be another geographer alive that has trekked the world as widely and as deeply as the Renaissance Man and geographer extraordinaire, Leon Yacher. Together, we've conducted fieldwork, exchanged views, swapped stories, and shared beer across the globe, often discussing matters pertaining to the history of geography and leadership in academe. The candid input he has provided over the years has been invaluable, especially with regard to this book. There is not enough that I can say to express my gratitude.

There is also an anonymous reviewer I must thank. His comments were much appreciated and helped make this a better book. Additionally, I must thank the late Van Cooley, Nancy Mansberger, Lou Ann Bierlein-Palmer, Andrea Beech, and Diane Bourgeois for insight vis-à-vis leadership in higher education today, Kathy Schoen for interlibrary loan assistance, and my Rowman and Littlefield editorial team, Alison Pavan, Emily Frazzette, and Ethan Feinstein. Qian Guo, Glenn Stracher and Mary Harris are owed a debt of gratitude for their unceasing support of more than two decades. Indeed, Matt West and Kearn Chouest must be acknowledged as well.

These acknowledgements would not be complete if I did not reflect on the impact of those that served in both formal and informal leadership roles during service in the armed forces years ago. G. Pereira was a role model for all in the Navy diving community. T.R. Woolens and D. Ortuno were exemplary military leaders as Recon Marines. It's been nearly forty years since my comrades, R.V. Rodriguez and L.G. Marinacci, walked with me among the elite of the elite in the service of Corps and country, and our very close friendship has persisted through the years. For that I am thankful.

Finally, I am grateful to my loving family. My mother and father are much admired for their parenting, perseverance, and guidance. My mother has been a stalwart example of courage, class, and character, and my late father epitomized transformational leadership. My children, Gabriel, Amelia, and Jesse did more than perhaps they ever will realize. If not for them, I wouldn't have chosen an academic life, nor would I have remained in academe. Moreover, no words are adequate to express my appreciation for my dear and loving wife, Holly. I am indeed fortunate and blessed to have the most beautiful woman in the world by my side.

ONE

Introduction

The evolution of any major discipline may be assessed by examining the emergence and decline of its academic programs *and its leadership* over the course of time. In the United States, widespread departmentalization occurred throughout academia in the latter part of the nineteenth century, and the discipline of *geography*, spawned by geology, most often found a home in academic departments devoted to the earth sciences. As the twentieth century unfolded, geography departments frequently gained autonomy and more readily incorporated the study of human activities within their curricula. For example, the discipline was included as an integral subject of study within the Department of Geology at the University of Chicago when the institution was founded in 1891; but not until 1903 was its Department of Geography established, which surfaced as the discipline's first independent graduate program in North America (Martin 2005). This launched a trajectory that several higher education institutions throughout the United States emulated; by 1925, Clark, Wisconsin, Columbia, Harvard, Berkeley, Michigan, and Yale received recognition as credible geography graduate programs (Warf and Webster 1980), and the discipline's exclusive professional organization, the Association of American Geographers, provided a network and annual forum for leaders in the discipline to emerge (James and Martin 1978).

Although it interrupted many graduate students in their quest for the Ph.D., World War II facilitated geography's prosperity, as the government's need for geographical information was dire and many geographers contributed to the war effort, among other things, in the analysis of intelligence (Barnes 2006). Moreover, in the War's aftermath, American colleges and universities swelled with students, many of whom who had seen service in European or Pacific theaters of operation; the need for qualified faculty in a number of fields, geography included, was great, as veterans of the armed forces enjoyed educational benefits in the form of the GI Bill. Often it was the academic leaders, the chairmen of graduate programs, who determined the course of the discipline's path toward success—or failure.

As enrollment increased new graduate programs emerged, such as the one at Penn State, and several departments that had already been established grew (e.g., Ohio State, Minnesota, and Syracuse). Here, transformational leaders oversaw the success of geography programs during this post-war era in which the discipline had already demonstrated its value to society. By the same token, laissez-faire leaders appeared to remain oblivious as other departments faded away; such was the case at Harvard at mid-century.

Thus, departments came and went over the remaining half of the twentieth century, and particular courses of study were introduced while others were removed; paradigms shifted, luminaries passed on, and their replacements established their own special interests. Concomitantly, personalities, politics, and prestige continued to play significant roles in the evolution of certain programs and the devolution of others. The discipline's history became marked by sagas that often read like soap operas, and students of the history of geography came to ponder these matters and draw conclusions from analysis of academic philosophies, conjecture, elements of the historical record, and hearsay evidence.

Though rich histories of geography departments and their leaders have been documented in a number of works, such as Martin's prosopographies (1968; 1973; 1980) and departmental histories (e.g., Trewartha 1978; Miller 1992; Clarke and Baumgart 2004), the emergence and decline of the programs that shaped the path the discipline has trod has not been addressed within the context of the theory and practice of leadership. It is argued here that leadership, particularly in reference to departmental chairmen, has played, and continues to play, a crucial role in the development of reputable geography departments. Given that the matter of leadership has not been addressed adequately in the geographical scholarship relating to the discipline's history, its examination is imperative; historians of the discipline, as well as its administrative leaders, stand to benefit much by moving beyond biographical studies and philosophical debates, and delving into the study of leadership.

It was James McGregor Burns' (1978) seminal work that introduced leadership concepts and attributes, which have since been examined by a number of students.[1] For example, Bass (1985) elaborated on Burns' definition of a transformational leader as "one who motivates us to do more than we originally expected to do" (p. 20), and at the opposite end of the spectrum, a laissez-faire leader as one who engages in "management by exception." Here, it is seen through historical analysis that strong leaders, often characterized as "transformational," established paradigms within their own programs and fueled the success of many faculty members and students. Conversely, ineffectual individuals miscast in leadership roles have far too often been found wandering at the helm of floundering programs, at times producing calamitous results.

The role leadership has played in the administration of geography departments has not been examined in reference to individual departments' accomplishments, and this work provides such a study to determine whether or not the leadership attributes of departmental chairmen in the historic past have influenced the success of their programs. It is argued that the types of leadership employed by chairpersons have been consequential to the achievement of departmental goals. In essence, certain leadership attributes have shown evidence of success in geography programs, and it is suggested that leaders characterized by those qualities that have tended to fuel successful programs should be emulated. Conversely, those characteristics that have been associated with leaders that have exhibited failure should be avoided.

Certainly, a balanced analysis of leading departments is essential to understand why some have flourished and others came to demise, for it is argued that it was not at all the esoteric nature of the discipline that sparked failings in academic departments, nor was it its eclecticism that generated success. On the contrary, throughout its history, the vitality of academic geography in the United States has rested upon its legacy of leadership. Essentially, effective leaders within key academic departments played imperative roles in the discipline's prosperity, and in contrast, mediocrity in leadership has largely been responsible for periods of austerity. Moreover, exemplary leaders characterized by transformational leadership attributes fostered high levels of achievement among their colleagues and subordinate faculty, whereas leaders marked by laissez faire characteristics often coexisted with unsuccessful colleagues. Thus, an examination of leadership in American academic geography, not solely relying on rumor and innuendo, but based upon historical analysis is imperative; for it reveals an understanding of the right moves and serious blunders that have created present circumstances. It is the intention of this work to offer such a perspective, so that leaders in geography, today and in the future, might be able to avoid the agony of failure and instead develop strategies for success by recognizing effective leadership behaviors that foster high levels of achievement.

The success of an institution very well may rest upon the shoulders of its departmental chairs (Arcuri 2002; Bennett 1988), and here it is maintained that because academic departments play an imperative role in the governance of American institutions, historical analysis is an essentiality, even though their contributions to the development of academic disciplines sometimes have been considered debatable. Kay Anderson (1977) commented,

> No administrative unit within the college or university has been so important, misunderstood, and maligned as the academic department. Prominent observers of the American scene have both castigated it for fragmenting higher education and extolled it for developing new knowledge. All of them, however, have agreed that departmentalization has played a dominant role in the evolution of higher education (p. 1).

This work supports this contention and also accentuates the role of departments in the development of a discipline beyond the contributions of individuals. Brown (1982) maintains that the nature of a department is reflective in the manner in which its members pursue group goals, which are usually secondary to the academic's individual goals. Seemingly, when the individual goals are considered a higher priority and incompatible with the departmental goals, the threat of departmental demise becomes real. Conversely, when the departmental goals become the goals of each of its members, then departments' maximize their opportunities for prosperity.[2]

Some students have claimed that the success of a university is based upon department chairs' effectiveness in leadership (Arcuri 2002; Bennett 1988). Certainly, the role of an effective chairperson is imperative in achieving success, and in academic geography, transformational leadership has been an effective strategy for herding cats and fueling the success of academic departments. How this leadership model has been successfully employed in the administration of geography programs has heretofore not been noted, and by developing an understanding of its past applications, much can be learned about the practice of transformational leadership, which can be emulated by leaders in the discipline today. Essentially, it is argued that the adoption of effective transformational leadership behaviors is the key to the successful chairmanship of academic departments.[3]

Although students of the history of the discipline have examined individual programs within the historical context, heretofore, no composite sketch has been made of the collection of American geography departments that dictated the discipline's trajectory over the last century; nor has an effort been made to synthesize the history of the discipline's most influential departments with critical examinations of their leaders vis-à-vis the study of leadership. This has been accomplished here through a thorough examination of the history of geographic thought in North America with particular reference to the leading academic programs.

It is partly through this type of historical assessment that Burns (1978) defined transactional and transformational avenues of leadership; yet, despite its viability, historical analysis has been noticeably absent in subsequent organizational studies. It is time to revisit the past by synthesizing studies in the manner of Burns and examining important organizational leaders that seriously influenced the shaping of the geographic discipline, assessing their leadership styles, determining their effectiveness, and perhaps establishing a model for examining leadership in other academic disciplines.

Granted, within the context of geography, the historical method has been employed, but concepts and theories relating to leadership in human organizations have been ignored. Weaving together the history of geography with leadership is an unexplored imperative, as it lends tremendous insight into the shaping of the discipline, for here it is maintained that its leaders throughout its history, largely at the departmental level, determined its course. This historical analysis offers a general perspective on the development of the discipline's leading geography programs throughout the twentieth century, and it identifies the leadership behaviors of historical figures associated with the geography departments that most influenced the development of the discipline.

Historical analysis of leadership in American academic geography reveals much about the prominent figures that influenced the course of the discipline, and this historical narrative provides a descriptive rendition of the leaders in terms of their backgrounds, talents, abilities, and will to achieve desired goals. Granted, in some cases insufficient evidence exists to verify dominant leadership behaviors; however, ample evidence exists to determine whether or not individuals exhibited a preponderance of transformational leadership attributes, and those that have demonstrated these behaviors have been identified as the transformational leaders in chairmanship positions that have influenced American academic geography.

Burns' (1978) identification of transformational leadership rests upon the premise that relationships are established for the achievement of a common purpose, which sparks a quest for higher values. The moral imperative is an essentiality, as is the selflessness of the leader. Transformational leaders are tasked with an *idealized influence* that is envisioning and worthy of emulation. Whereas "transactional leadership models are grounded in a worldview of self-interest . . . authentic transformational leadership provides a more reasonable and realistic concept of self—a self that is connected to friends, family, and community whose welfare may be more important to oneself than one's own" (Bass and Steidlmeier 1999, pp. 185–186). In this particular case, the academic department is the family, the institution is the community, and society is the community expanded.

Transformational leaders are tasked with engaging in *inspirational motivation,* which enables followers to greet challenges with zeal and realize the value of the journey toward achieving the dream. Here the dream is to achieve academic excellence; with the conquest of knowledge and attainment of goals comes a level of empowerment. *Intellectual stimulation* also plays an important role in transformational leadership in that the leader seeks to establish dialogue about the particular merits of issues with no hidden agenda attached. Here, the search for truth and the attainment of sound scholarship fulfills this goal. Lastly, *individualized consideration* as an essential component of transformational leadership is oriented around the mentoring of followers to achieve success and to become leaders themselves (Bass and Steidlmeier 1999).

The description of transformational leadership with special reference to its application in academic geography is enhanced by noting five principal attributes that characterize stellar transformational leaders of the recent historic past: *inspiration, integrity, selflessness, scholarship,* and *proaction*. The leader provides *inspiration* through a variety of means, and often this incorporates a magnetic charisma, which attracts followers, whether they are faculty or students, to serve with the leader. The inspiration may come from a legacy of success in scholarly achievement, raising funds for faculty and students, or studies in the classroom or in the field that spark a desire to carry out a life of learning as a geographer. It is here that the follower realizes that the leader is a model to follow, and it is here that the leader essentially articulates a vision that followers seek to embrace and pursue. *Integrity* is an imperative to maintain moral authority. The office of leader requires maintaining honest relations with colleagues, subordinates, and students, because trust is essential and credibility is lost without high moral standing.[4] The transformational leader is compelled to act with *selflessness*, insuring that the needs of the followers are met as much as possible, while also addressing the needs of the greater community. The leader must consider himself (or herself) as one member of a cohesive unit, albeit the one member with authority and responsibility.[5] Often, assisting in achieving the individual goals of fellow faculty and students takes precedence over time spent in pursuit of personal goals. As a role model, the transformational leader must engage in *scholarship*. He/she must continue to conduct credible research and publish the related results while also serving as an administrative leader. *Proaction* is a requisite form of behavior, rather than reaction to a specific issue that may have ramifications for particular individuals, whether they are students or faculty, or the department (or program) as a whole. Essentially, the transformational leader is charged with having the insight to engage in proactive intervention before destructive threats to individual colleagues, students, or the program become real. This often requires building relationships and fostering a network of trusted allies throughout the university community.

It has been determined which departments and programs have stood head and shoulders above other geography departments, creating the most significant impact in shaping the geographic discipline. Reflection on the discipline's history reveals that before national rankings became important, relatively few programs offered doctorates before the end of World War II; however, it was those graduate programs that created a body of alumni who played major roles in the development of geography. Sixteen doctoral programs have been identified as being markedly influential in shaping the course of the discipline; all but two were ranked among the "Top 10" geography programs in the United States at least once since the Second World War, and since then, four have folded (Harvard, Michigan, Northwestern, Chicago) and only one (Santa Barbara) was not in existence before the 1980s. A remarkable level of consistency has characterized the status of these departments; in five national rankings between 1957 and 1995; five programs appeared in the "Top 10" every time (Berkeley, Wisconsin, Minnesota, Washington, UCLA) and three other programs have been ranked in the "Top 10" four times (Syracuse, Chicago, Ohio State). The sixteen programs listed in Table 1.1 have been examined through historical analysis in order to develop an understanding of the leadership attributes that influenced various levels of achievement.

Table 1.1. Geography Doctoral Programs Shaping the Discipline's History

1	University of Chicago (*Chicago*)
2	University of Wisconsin (*Wisconsin* or *Madison*)
3	University of Michigan (*Michigan* or *Ann Arbor*)
4	University of Minnesota (*Minnesota*)
5	Ohio State University (*Ohio State*)
6	Pennsylvania State University (*Penn State*)
7	Syracuse University (*Syracuse*)
8	Clark University (*Clark*)
9	Johns Hopkins University (*Johns Hopkins* or *Hopkins*)
10	Harvard University (*Harvard*)
11	Northwestern University (*Northwestern*)
12	University of Washington (*Washington*)
13	University of Iowa (*Iowa*)
14	University of California, Los Angeles (*UCLA*)
15	University of California, Berkeley (*Berkeley*)
16	University of California, Santa Barbara (*Santa Barbara*)

Note: Name in parentheses (e.g., UCLA) refers to the program's vernacular identification. Numerical ranking is *not* intended to connote department's level of significance.

Although it can certainly be contentious an examination of national rankings of graduate programs is essential. The 1995 NRC report is dated; however, it is a standard that has gained widespread acceptance despite the release in September 2010 of the National Research Council's newest rankings (Castree et al. 2006; Doctoral 2010). The data accumulated for the 2010 assessment have been brought into question and the methodology employed makes program assessment nearly impossible.[6] Granted, since the 1995 study took place, new doctoral programs have emerged, and some shuffling (and reshuffling) in the rankings has occurred; however, regardless of any reordering that might have taken place, the programs listed in Table 1.2 continue to be widely considered as among the leading departments of geography in the United States, and instead of being assessed in rank order according to the 1995 report, grouping them as peers in a single category is a more reasonable approach to consider in this current study.

Table 1.2. Top 10 Geography Doctoral Programs, 1995

1	Penn State
2	Wisconsin
3	Minnesota
4	Santa Barbara
5	Ohio State
6	Syracuse
7	Berkeley
8	UCLA
9	Clark
10	Washington

Source: Goldberger 1995.

This work relies on the many pieces of the discipline's history that have been recorded by historians of geography as a foundation for historical analysis. Published works on the histories of various programs, biographies of selected leaders, and filmed interviews have been integrated with recollections and notes of my own dialogues with luminary geographers over the course of nearly three decades. These sources, when woven together, reveal not only the historical development of the programs that shaped the course of the discipline, but also the leadership attributes of those that served as chairs and program directors. The departmental leaders are assessed based upon their leadership behaviors, and through this assessment, various levels of program success are attributed to the leaders.

Given that this study is devoted to examining the impact of the most influential departments in the history of geography in the United States, some students in the field might argue that the decision to include or exclude individual departments has been arbitrary. The rationale for inclusion of departments has been based on the evaluation of departmental rankings that have been conducted over the course of more than fifty years, and the inclusion of additionally selected individual programs not ranked in the "Top 10" in any survey, Johns Hopkins and Iowa, is based upon the important role these departments played in the development of geography as a discipline. By the same token, three departments that entered the "Top 10" in 1957, 1964, and 1969 are excluded from the study; the University of Illinois, Louisiana State University (LSU), and the University of Kansas despite their significant contributions, did not create an impact on par with other departments examined.[7]

In any event, of significance to this study is the National Research Council (Goldberger 1995) report on Ph.D. programs in the United States. Of the fifty-three geography doctoral programs that awarded Ph.D.s, thirty-six were included in the study, which accounted for 84 percent of all doctorates awarded between 1986 and 1992. Programs were rated based upon the scholarly quality of program faculty and the program effectiveness in educating research scholars and scientists. With the exception of Santa Barbara, which was not included in the earlier 1983 NRC program assessment, all programs that ranked in the "Top 10" in the 1995 report had been ranked in the "Top 10" in 1983; however, perceived declines in program quality were noted for the departments at Minnesota, Ohio State, and Berkeley.

Previous national rankings also prove to be useful in establishing long-lasting legacies for individual programs and indicating the fall of others. Table 1.3 illustrates the "Top 10" geography programs in five different national rankings conducted between 1957 and 1995. In an unpublished work, Warf and Webster (1980) examined geography program rankings between 1925 and 1980, which include assessments by non-geographers and geographers alike. Additionally, the National Research Council's 1982 study (Jones et al. 1982) provides a solid basis for the rankings of academic departments after the Second World War, as does the Cartter report (1966), the Keniston study (1957), and the Roose (1970) assessment. Moreover, Berelson's 1960 study (2008) documented the evolution of graduate education in the United States. For example, he noted that until the 1920s, Columbia, Chicago, Harvard, Johns Hopkins, and Yale awarded approximately half the doctorates awarded in the U.S.; a decade later, the five most productive institutions awarded only a third of American Ph.D.s, and this list included: Columbia, Chicago, Harvard, Wisconsin, and Cornell. By the 1950s, Columbia, Wisconsin, Berkeley, Harvard, and Illinois awarded less than 25 percent of American doctorates as the most productive five institutions.

Table 1.3. Top 10 Geography Graduate Programs, 1957–1995

Rank	1957	1964	1969	1982	1995
1	Wisconsin	Wisconsin	Chicago	Minnesota	Penn State

2	Chicago	Chicago	Michigan	Penn State	Wisconsin
3	Berkeley	Berkeley	Minnesota (3)	Berkeley	Minnesota
4	Michigan	Washington (3)	Wisconsin (3)	Chicago	Santa Barbara
5	Northwestern (4)	Syracuse	Berkeley (3)	Wisconsin	Ohio State
6	UCLA	Northwestern	Washington	Ohio State	Syracuse
7	Washington	Minnesota	Ohio State	UCLA	Berkeley
8	Ohio State	UCLA	Penn State	Clark	UCLA
9	Illinois	Michigan (8)	Syracuse	Washington	Clark
10	Minnesota	LSU	Kansas (10)	Syracuse	Washington
11			Northwestern (10)		

Sources: Keniston 1957; Cartter 1966; Roose 1970; Jones et al. 1985; Goldberger 1995. Numbers in parentheses indicate placement of two or more programs that are tied in the national rankings.

Fortunately, as Kenzer (2001) has noted, "Unlike practitioners in other academic fields, when it comes to the intellectual history of our discipline, geographers love to dabble" (p. 752). The field is rich, and a selection of sources is found in the bibliographic essay in Appendix B. Suffice it to say that the departments of graduate programs in geography and the leaders that have shaped the course of the discipline have been examined. Only the sixteen programs that *most* influenced the field of geography constitute the subject of this work, and given our predisposition to "dabble" in the historic past, as time goes on, more and more evidence will be accumulated documenting and analyzing the history of American academic geography. As scholarship is produced, it is likely this work will raise questions and spark debate.

Regardless of other perspectives that arise, the story that unfolds in the following pages is my account of the geographic discipline's history in the United States with reference to its most significant leaders during the twentieth century. Here, context is important, and Martin's (1985) paradigms in American geographic thought offer a viable model for framing this narrative. Given that it was a fledgling discipline at the dawn of the century, a brief overview of the discipline's early years during a paradigm marked by *physiography and causation* is offered in a single chapter. Detailed examinations of leadership in each department during the *field and region* paradigm are offered in subsequent separate chapters, for it was during this time that nearly all of the programs developed some element of distinction. *Eclectic pluralism* characterizes the following paradigm, and a single chapter is devoted to discussing the principal leadership attributes of each department. After this chapter is one followed by discussion of the emergence of a new paradigm, *Geographic Information Science*, and the creation of a new department of geography. The concluding chapters offer insight and perspective, as well as suggestions for future research. In essence, the intention of this work is to initiate dialogue on leadership that will advance the discipline and spark transformational leaders to emerge.[8]

NOTES

1. The dean of leadership scholarship, author of numerous books, Woodrow Wilson Professor of Government Emeritus at Williams College and winner of the National Book Award and Pulitzer Proze for *Roosevelt: The Soldier of Freedom*, James McGregor Burns passed away in July of this year at the age of ninety-five. He revolutionized the study of leadership by focusing on the phenomenon as a relationship between the leader and the led. Subsequently, a number of students, perhaps most notably, the late Bernard Bass, elaborated on the work of Burns in efforts to identify the attributes that corresponded with transformational leadership behaviors.
2. The success experienced by individual geography departments in the discipline's history, such as Berkeley under Sauer and Santa Barbara under Simonett, support this contention. A common vision works wonders.
3. Appendix A offers a synopsis of research devoted to leadership in academic departments.
4. In no case should academic dishonesty to be tolerated. Yet, it occurs far too frequently in graduate programs. For example, there are a number of cases in which faculty have taken the scholarship produced by students and presented it as their own. It is unfortunate that the academic community of geographers finds this tolerable.
5. Workloads and salaries should be assigned as equitably as possible, committee work should be shared, and students should be given timely guidance and direction in the completion of their studies.
6. The 2010 NRC report, for the purpose of this work, is not considered a viable means of assessment. Although long awaited, the National Research Council (NRC) report released in September 2010 (Doctoral 2010) is perhaps the least useful assessment. It is the newest ranking for doctoral programs in the United States and marks the first assessment of these programs since the 1995 report was published. In order to evaluate program quality, the NRC asked faculty about different variables used to assess programs and weighted them based upon the degree to which faculty believed them to be important. This methodology employed was significantly different than that used in the 1995 study; instead of specific values and rankings, programs were accorded a range. With regard to geography programs, ranges spanned up to twenty-four places; for example, such is the case with Texas A&M University's program, which is ranked twenty to forty-four out of the forty-nine total programs surveyed. This wide level of variance not only makes the 2010 assessment unwieldy and difficult to use for ranking programs in terms of quality, it presents, at the very least, questionable conclusions.
7. Rankings did not take place before Harvard was terminated shortly after World War II; otherwise it would have made the list of "Top 10" programs simply because of the dearth of geography graduate programs. On the other hand, it is in a sense surprising that the University of Iowa's department, certainly one of the more significant players in the discipline's history, never entered the "Top 10." Here, what might be considered an intangible criterion adds Iowa to the list of the most influential departments, and it relates to the founding of a new paradigm in the 1950s; Iowa's influence

far surpassed any other second-tier geography graduate program, and excluding it from this list would render an incomplete account of leadership in the history of geography. Additionally, Johns Hopkins is not on the 1995 list; however, here, too intangible factors played an important role in its contributions to the development of the discipline's history, and this related to the influence of the president of the institution, who happened to be a geographer of distinction engaged in advising U.S. presidents, in addition to a national scandal that erupted at Hopkins in the geography program, which was associated with the McCarthy witch hunts.

8. Currently, about fifty-eight graduate American universities offer doctorates in geography. Although a relatively recent emergence of Ph.D. programs has occurred at a number of public universities (University of Connecticut, University of South Carolina, City University of New York, University of North Carolina-Charlotte, George Mason University, University of Nevada-Reno, University of Southern Mississippi, Kansas State University, University of South Florida, and Oklahoma State University), these have not been enough to make up for the loss of departments in relatively prestigious institutions, which has occurred since the end of the Second World War (Harvard, Yale, Columbia, Chicago, Northwestern, Pittsburgh, Michigan, and the University of Southern California). Today, there are no geography Ph.D. programs in U.S. News and World Report's (2014) top ten national universities (http://colleges.usnews.rankingsandreviews.com/best-colleges/rankings/national-universities). Dartmouth is tied at number ten, but it offers only the geography baccalaureate. Although Johns Hopkins ranks number twelve, its combined geography and environmental engineering department with only one faculty member in geography shows very little promise for advancing the geographic discipline. In essence, the only true geography doctoral programs listed among the top thirty nationally ranked universities are: Berkeley (tied at number twenty), UCLA (tied at number twenty-three), and the University of North Carolina at Chapel Hill (number thirty). As geography struggles (and sometimes succeeds) in advancing its case for legitimacy in American higher education, it must be acknowledged that with only three Ph.D programs in the top thirty, the challenge will be formidable. It will be one for the most exemplary transformational leaders in our discipline to take on.

TWO

The Ascent of American Academic Geography

Before its formalization in 1892, geography had been practiced in North America by a number of scholars such as Ben Franklin, Thomas Jefferson, and George Perkins Marsh. Field surveys devoted to exploration of the American West during the nineteenth century conducted by Meriwether Lewis and William Clark (1803–1806), Ferdinand Hayden (1869–1878), and John Wesley Powell (1867–1880) also contributed much to understanding about the people and places that made up previously uncharted territory (Martin 2005).

By the end of the Revolutionary War, Cornell was the only Ivy League institution that had not yet opened, and all, in addition to William and Mary and Rutgers, taught geography as integral to the standard course of study. The subject incorporated the use of sundials, globes, survey and navigational instruments, and maps to foster an understanding of the earth. It also included textbooks and maps, which appear largely to have been devoted to the location of geographical phenomena (Simpson and Huke 1988). As an academic field, geography was highly regarded. Statesman and orator Daniel Webster wrote to his nephew at Dartmouth in 1814, "Geography is an interesting study. It is an indispensable preliminary to history . . . I would advise you never to read the history of any country till you have studied its Geography" (Simpson and Huke 1988, p. 11).

As American higher education evolved during the nineteenth century, geography as a discipline was usually included in college curricula. In the Ivies, the subject was discarded for a short time during the antebellum period, for it appears that college preparatory schools filled the need in providing geographical instruction. Incoming students were expected to have mastered the fundamentals of geographical knowledge before beginning study at the most stellar institutions (Simpson and Huke 1988; Wright and Koch). Elsewhere, the situation was different. A number of courses, such as *ancient geography, physical geography,* and *modern geography,* were taught before the Civil War at the antecedents to the land grant universities in Pennsylvania, Tennessee, Alabama, and Missouri (Blouet and Stitcher 1981; DeVivo 1990). And Swiss-born Arnold Guyot brought the teachings of the German geographers, Alexander von Humboldt and Carl Ritter, to Princeton where he held the position of professor of physical geography and geology between 1854 and 1880 (Martin 2005). This is notable because although a number of Ivies appeared to have discarded geography as a core subject in the mid antebellum period, it seems to have been resuscitated as an important physical science in the middle of the nineteenth century.

When Daniel Coit Gilman became president of Johns Hopkins University in 1876, the idea of a university as "a society of scholars" emerged and diffused throughout American academe (Martin 2005, p. 370). The notion of well-educated professors that continued to engage in research, chosen to teach bright young minds, developed as a trend, and the professionalization of disciplines occurred. This was the type of university that had recently arisen in Germany where geographers played a crucial role, and American institutions followed suit. Courses of geography instruction quickly expanded and departmentalization occurred, though in American bellwether universities geography met with variable levels of success (Dunbar 1986).

During the paradigm, *physiography and causation* (Martin 1985), no one did more than the eminent geomorphologist at Harvard, William Morris Davis, to spark the teaching of geography in American higher education at this time. He was known as a particularly captivating lecturer with a devotion to research and leadership in the geographic discipline that was unmatched (Beckinsale 1981; Hartshorne 1981; Martin 2005). Davis also played an important role in developing the Committee of Ten's guidelines for geographic education. It was under the supervision of T.C. Chamberlin, formerly president of the University of Wisconsin and now chair of geology at the University of Chicago, that the Conference on Geography for the Committee of Ten, which was established by the National Education Association in 1892 to examine college preparation and college entrance requirements, recommended vast changes to improve the pre-college geography curriculum. Geography was then formalized as an academic discipline; shortly thereafter came the emergence of geography departments and programs in flagship institutions, normal schools, private liberal arts colleges, and elsewhere (Martin 2005).

Though the University of California at Berkeley claimed it established the first full-fledged department of geography, Harvard, Yale, Pennsylvania, and Chicago proved to be the most important grounds for the training of academic geographers and promotion of ideas before the

1920s. Davisian disciples trained at Harvard went on to promulgate geography education at various institutions ranging from Ivy League Columbia to the normal school in Ypsilanti, Michigan. Herbert Gregory, who worked with Indians in the Southwest, finding them water, chaired the geology and geography department at Yale in which he assembled on the staff a number of capable geographers: Hiram Bingham, who lectured on Latin American geography and rediscovered Machu Picchu in 1911, as well as the distinguished climatologist, Ellsworth Huntington, and the later advisor to two U.S. presidents, Isaiah Bowman, among others (Martin 1968; Martin 1973; Martin, 1980; Martin 2005). Martin (1997) maintained that Yale produced "some of the earliest legitimately geographical doctorates in the country" (p. 18). At the University of Pennsylvania's Wharton School of Finance and Commerce, Emory Johnson in 1893 had completed the first American Ph.D. dissertation in geography on "Inland Waterways: Their Relation to Transportation," and there he developed a sound geography curriculum, serving the institution for forty years (his last four as Dean). Incidentally, in 1899, he was commissioned by the U.S. Congress to investigate the economic feasibility of a canal across the Isthmus of Panama and was joined by a graduate student named J. Russell Smith who later became a luminary in his own right (James and Martin 1973; Martin 2005). At the University of Chicago, established in 1891 as a leading institution of exemplary scholars, notable geologists such as T.C. Chamberlin and Rollin Salisbury were immediately hired, and a separate geography department was established in 1903 devoted largely to graduate work (James and Martin 1973; Martin 1985; Martin 2005). Within a quarter century, Chicago Ph.D.s were teaching at Michigan, Wisconsin, Harvard, and Clark, as well as Chicago itself. Until 1923, it was the only school where one could be awarded a geography doctorate from a department of geography, and there the role of Salisbury cannot be underplayed. Seventeen years after his death in 1939, more members of the Association of American Geographers reported that Salisbury influenced them more than any other geographer (Bushong 1981; Pattison 1981; Rugg 1981).

Essentially, the actors who played leading roles in geography during the pre-World War I period were instrumental in launching geography as a credible academic discipline. Carl Sauer, a student of Salisbury, would later refer to this time as "a springtime, the only good one we've had in this country" (Martin 1985, p. 232). Martin (1985) identified the paradigm in American geographic thought that prevailed between 1892 and 1925 as being largely oriented around the notion of environmental determinism, which in its rawest sense referred to the idea that the physical environment dictated human activity. Here came an important departure from geology, which concentrated on the natural processes that shaped the earth with little concern for the role of man.

Given geography's lineage in North America, many geographers at the time were skilled in geomorphology and sought to explain differences and similarities among people and places by accentuating the causal relationship between the elements of physical geography and humans. Davis had been a catalyst in this movement, and he developed a strong following among his students as well as colleagues in the newly formed Association of American Geographers. Among them were Albert Perry Brigham who taught at Colgate and authored a widely used textbook, *Geographic Influences on American History*; Ellen Churchill Semple who lectured at Chicago and Clark and also authored a similar text, *American History and Its Geographic Conditions*; and Ellsworth Huntington who served as an assistant professor and research associate at Yale and wrote a number of books on the subject including: *Civilization and Climate*, *Mainsprings of Civilization*, and *The Character of Races as Influenced by Physical Environment, Natural Selection, and Historical Development*, (Martin 1973; Martin 2005). The influence of these writings was pervasive; Huntington's works remained in print throughout most of the twentieth century and a number of institutions apparently adjusted their geography curricula to include courses on *North America* that were named after the texts authored by Brigham and Semple.

Elaboration of environmental determinism is indeed in order, for it typified the paradigm that prevailed from the founding of the academic discipline in North America to its early adolescence after the First World War, and here some discussion of its major players and their influences is apropos. William Morris Davis, the Harvard geologist, coined a new term, "ontography," which was oriented around the concept of human response to environmental conditions, and he sought to have this maintained as a guiding principle for the study of geography in the pre-college curriculum. His doctrine dictated that the environment controlled human activities, which largely disregarded the influences of culture. Yet, his premise found fertile ground for research by a number of geographers. Ellen Churchill Semple explored the mountains of Kentucky and in appealing prose noted that topographic barriers facilitated an isolation, which preserved unchanged traditions and cultural elements such as the English dialect spoken from the Elizabethan era (Semple 1910). Ellsworth Huntington conducted statistical analyses to determine optimal temperatures for both mental and physical work, and he correlated these examinations with assessments of climatic controls and economic productivity (Martin 1973).

This is not to say that all geographers embraced determinism. Notables like Mark Jefferson, Isaiah Bowman, and Rollin Salisbury were disinclined to accept the causal notion. Salisbury appeared more interested in the impact of humans on the environment than on environmental influences on human activities, and he cultivated this manner of thinking among his students, including Carl Sauer. Sauer appeared to envision a more holistic approach to studying geography by examining regions, which he emphasized while on the faculty at the University of Michigan during World War I. There he offered a regional course on the *Geography of North America*, which was offered in a number of other institutions as well (Martin 1981). Certainly, geography was not only vibrant in northeastern colleges and universities; it was gaining considerable ground in the Midwest.

Unfortunately, however, geography in the pre-college school curriculum was meeting its demise, despite the success of a new professional organization devoted to schoolteachers of the discipline. Apparently, the recommendations made by the Committee of Ten were not feasible, as teachers were not usually properly trained in the normal schools to teach the field adequately. Consequently, by 1912, geography was considered "a dry, uninteresting subject" and relegated to a minor role subordinate to history and civics in the recently created social studies (James 1971, p. 377).

In any event, by 1925, the discipline's status had declined considerably in the Northeast, and some have speculated that this was because university administrations and colleagues in other departments viewed causation as overly simplistic; thus, environmental determinism fell out of favor. Though this conclusion seems sensible, it is not supported by any reasonable evidence. Granted, as Erhard Rostlund was to comment about this time three decades later, "Environmentalism was not disproved, only disapproved" (Rostlund 1956, p. 23), and the causal notion was viewed rather unfavorably; however, to suggest that the discipline was held in low regard because of a paradigm that was waning in popularity, is purely a speculative jump to conclusions without considering other factors. Indeed, other circumstances contributed to the discipline's decline, and these matters relate to leadership changes in the field, particularly in geography programs in higher education institutions.

In certain cases, it appears that because geography was subordinate to geology in binomial departments, it failed to win battles for funding. At Yale, for example, Gregory became ill, a geologist took over as chair, and "the University instituted economies which decimated the geographers" (Martin 1997, p. 18). At Harvard, Davis retired, leaving no one willing to serve as a staunch advocate for the discipline.[1] Additionally, it should be noted that by the 1920s, a large number of physical geographers who had served as the driving force of the discipline in its formative years died within a very short period (DeVivo 1987; Wright and Koch). For example, Ralph S. Tarr, who authored leading physical geography textbooks at the time, was doing much to create a credible geography program at Cornell when he unexpectedly passed away in 1912. Eminent geomorphologist Grove Karl Gilbert died in 1918 and Salisbury, also a prominent textbook author, died in 1922. A youthful discipline that had grown vibrantly from its geologic roots now saw its connection to earth science waning with the loss of giants in its leadership that had no comparable replacements.

Economic geographer Walter S. Tower had moved from Ivy League Penn to take a position at the University of Chicago in 1911. With less emphasis on the physical environment more attention was paid to economic geography, which lent itself quite nicely to the regional geography advanced at Michigan and Chicago. A "Midwest takeover" occurred as the discipline's younger practitioners had now successfully gained control from the older geographers who had been inspired by Davis (Martin 1985).

When Sauer left the University of Michigan to take the chair at Berkeley, he delivered an address espousing his ideas of geography along the line of *region,* not *causation* (Sauer 1925). This was published in 1925 and considered the crushing blow for environmental determinism and the beginning of a new paradigm, *field and region* (Martin 1985).

Geography as an academic discipline was expanding across the country between the end of both World Wars in normal schools, colleges, and universities; however, its position in the stellar Ivies had lost much ground to the public universities that had sprung up. This was particularly the case in California, Wisconsin, and Michigan, which fueled solid programs through collections of rare talents in faculty and students. Though the discipline survived at Harvard, it was overshadowed by geology in a binomial program, and its neighbor, Clark, with a geographer at the helm of the institution's presidency, emerged as the leading school of geography in the region. Moreover, with another geographer as president of Johns Hopkins, geography was incorporated as an integral component of the curriculum shortly after Pearl Harbor.

Yet the University of Chicago arguably continued to command the lead among American higher education institutions. The shift in the discipline's leadership from Harvard, where Davis had essentially sparked the beginning of academic geography, but failed to develop a solid program, had occurred. James and Martin (1978) remarked, "Transfer of control of the Association [of American Geographers], from the disciples of William Morris Davis to those of Rollin D. Salisbury, was quietly and certainly one of the major developments" (p. 69) of the period between the World Wars. Seemingly, the institution in Cambridge was hampered by an archaic administrative element that was slow to embrace changes, and the Midwestern institution was willing to accept innovations; conditions were ripe for change, as Salisbury stepped down as graduate school dean in order to chair the first doctoral-granting geography department in the United States (Rugg 1981).

NOTES

1. And his successors did not necessarily impress the Harvard administration in the areas of teaching or research (e.g., Atwood).

THREE

Transformational Leadership at Chicago

The Post-Salisbury Years

Geography in the Midwest continued as an enduring presence at Chicago, but with the passing of Salisbury, physical geography slowly faded from the curriculum. Chauncy Harris (1979), a distinguished alumnus (and faculty member), wrote in his brief history of the department that "In the thirties the graduate program within the Department of Geography at the University of Chicago seemed to me to be weak in physical geography, although students were encouraged to take work in geomorphology with J. Harlan Bretz in geology and work in plant geography with Henry C. Cowles in botany" (p. 29).

The role Cowles played was highly significant, for he was a founding member of both the Association of American Geographers, as well as the Ecological Society of America, and he served as president in each. Cowles' history with geography went back a number of years, and as he was educated at Chicago, it's important to note that of the thirteen graduate courses he took, twelve were in geology; this shows a strong Salisbury connection (Pattison 1981). As a de facto member of the geography graduate program faculty, he introduced students to ecological and biogeographical matters through lectures, as well as the fieldwork he conducted in the Great Lakes region.[1]

Other members of Chicago's faculty also made a conscious effort to write textbooks in their areas of expertise for students in higher education as well as in grades K–12. Moreover, by the 1930s, the department's graduates were serving on the faculties of sixty colleges and universities in addition to seventy normal schools and teachers colleges, which is arguably a more accurate measure than any for assessing the viability of a program (Harris 1979).

As the department's strengths were in the human side of geography and regional studies, students at Chicago were pointed in that direction, and newly hired faculty were often chosen for their regional specialties. J. Paul Goode, a Ph.D. in Economics from the University of Pennsylvania, taught at Chicago from the turn of the century until 1932; Harlan H. Barrows, who had earned a teaching certificate at the Normal School in Ypsilanti, Michigan, started his academic career at the Ferris Institute in Big Rapids, Michigan.[2] He went on to complete the baccalaureate at the University of Chicago, and there he was known as an exemplary student. Quickly finishing his graduate studies, he rose to the rank of professor only eleven years after completing the B.S. in geology, going on to serve as department chair from 1919 to 1942. Drawing influences from both the historian Frederick Jackson Turner and geographer Ellen Churchill Semple, Barrows crafted a very popular historical geography course to help explain the relationships between people and the environment in the United States (Koelsch 1969).[3] In fact, a survey of alumni reported that his historical geography course was one of "the most worthwhile courses in the whole undergraduate program of the university" (Martin 2005, p. 373). Barrows brought in a number of distinguished academic geographers from across the globe to lecture at what had quickly become one of the leading universities in the world (Colby and White 1961).

Thus, several international figures served on the faculty, and among them was Australian T. Griffith Taylor who taught at Chicago between 1928 and 1935 before moving on to build the department at the University of Toronto taking over as chair. Certainly, the leadership role played by Charles Colby worked in the program's favor. Like Barrows (and several other luminaries in American geography), Colby studied under Mark Jefferson at the Michigan State Normal School in Ypsilanti before completing the baccalaureate. Afterward, he joined the faculty at the Normal School in Winona, Minnesota, and later pursued graduate studies at Chicago, completing the Ph.D. in 1916. Immediately becoming a faculty member, Colby began his career in guiding many students through the completion of their doctorates. Edward Ullman claimed that Colby was especially inspiring (Dow 1972), and Chauncy Harris later wrote of him,

> His concern for students was revealed in the initiative he took in getting acquainted with them as whole persons, and in his discussions with them not only about geography, but also about their lives and their potential careers. He saw their potential qualities and emphasized these. For many students he opened up vistas of the potentialities of geography and stimulated and encouraged them actively to engage themselves in the exciting adventure of its exploration (Harris 1966, p. 380).

Contributing much in geographical scholarship, Colby engaged in applied studies devoted to urban geography and economic development, while also serving in higher education leadership positions including the role of department chair between 1942 and 1949 (Harper 2005; Mikesell 2004). As the mentor of several students, including two in urban geography that would later become luminaries in the field, Chauncy Harris and Edward Ullman (interestingly, both sons of highly regarded members of the professoriate), Colby led one of the best geography departments in North America.

At Chicago, fieldwork continued to carry on a strong tradition. The summer field courses were considered an imperative in geographical study, for restricting instruction to reading books and classroom discussions posed limitations on comprehending the nature of landscapes. Robert Platt, only the fourth chairman (1949–1956) in the fifty-plus-year history of the department was known as a particularly effective professor of the field course, at times leading students from the Straits of Mackinac to Hudson's Bay (Mikesell 2004). "He was ever the explorer: exploring ideas, places, and peoples" (Hartshorne 1964, p. 630). Platt had studied philosophy at Yale and spent a year in China; with no academic background in the subject, he was assigned to teach geography. Returning to the United States, he decided to pursue graduate studies in geography at Chicago, becoming an instructor in 1919 and rising to full professor twenty years later (Hartshorne 1964). Known for his generous hospitality, Platt and his wife opened their home to board a large number of students who would come "to feel themselves part of the family" (Hartshorne 1964, p. 631). More than 100 students from the U.S. and abroad stayed with the Platts for at least a year, and this indeed showed the dedication of the faculty to their students.

In essence, the scholarship and charisma that marked Salisbury, the magnanimity offered by Platt, the inspiration generated by Colby, and the dedication captured by Barrows simply highlight traits of the department's chairmen throughout its first half-century of existence. The leadership created not only a climate that facilitated sound geographical research, but also one that appeared to develop transformational leaders. These solid attributes seem to have diffused to their students that earned doctorates and served on faculties elsewhere as leaders in either formal or informal capacities.

Unfortunately for some, the university enforced a mandatory retirement age of sixty-five, and though Platt remained committed to conducting extended field studies, frequently in Latin America, he was compelled to relinquish the popular field course to younger members of the faculty; Harold Mayer and Allen Philbrick took it over (Hartshorne 1964; Mayer 1979; Mikesell 2004).[4] Still, although the Chicago department remained a powerhouse in the discipline during the post-World War II period, the mandatory retirement age of sixty-five resulted in losses; this was a factor in T. Griffith Taylor's decision to leave for Toronto, for there, retirement was mandated at seventy years of age, which gave him five more years to devote to the discipline (Rose 1964).

Age also played a role in other ways; emeritus faculty that exhibited transformational leadership attributes and retained some influence in departmental matters because of their luminary status passed away within a very short period shortly after the shift in paradigms occurred: Barrows in 1960, Platt in 1964, and Colby in 1965. Though still solid, Chicago would no longer hold a wide edge over other departments as the premier graduate program in North America, for public universities from the West Coast to the Midwest offered stiff competition, and Clark and Syracuse in the Northeast had emerged as stellar forces.

NOTES

1. In the summer of 1985, I visited Fred Kniffen, at Louisiana State University (one of Carl Sauer's earliest students). At eighty-five years of age, he maintained a strong presence, coming into his office in the combined geography and anthropology department daily, and he continued to be highly regarded as a cogent scholar that carried the wisdom of his years. Having developed some level of expertise in the history of geographic thought, I was excited to meet with him and he seemed happy to reminisce with me about the past. Among other things, I indicated to him that although I had seen nothing in print concerning the matter it appeared to me that Sauer had been influenced at Chicago just as much by Cowles as by Salisbury. Kniffen supported this contention and confirmed that while reflecting on his graduate studies, Sauer had remarked on the role Cowles played, indicating that the impact of the eminent ecologist was equivalent to the role played by Salisbury, if not more so.

2. Isaiah Bowman, the notable geographer who directed the American Geographical Society, served as president of Johns Hopkins University, and advised two U.S. presidents, was a student of Barrows in Big Rapids.

3. Over the course of time, Henry Chandler Cowles' influence on Barrows proved to be pervasive. This was evident in Barrows presidential address to the Association of American Geographers, "Geography as Human Ecology."

4. Mayer contributed much to scholarship in urban geography, in which the department had begun to develop a reputation (Dow 1971).

FOUR

Paradoxical Leadership at Wisconsin

In 1892, the president of the University of Wisconsin, T.C. Chamberlin, resigned to accept the chairmanship of the new Department of Geology at Chicago, taking with him the prominent geomorphologist, Rollin Salisbury, who had been teaching a course largely devoted to physical geography in the geology department on the Madison campus. Some courses in economic and commercial geography were offered in the School of Economics, Political Science, and History; however, geography shared a strong presence in the geology department, largely due to the influence of the notable physical geographer, Nevin Fenneman, who had been educated at the University of Chicago. Unlike geographers of that time who embraced causation, Fenneman argued that "the task of geography was to grasp the content and individuality of areas" (Trewartha 1978, p. 7). Arriving in Madison in 1903 and departing only four years later to begin a new program at the University of Cincinnati, geography enrollments increased exponentially during his brief tenure, and the mark he left on colleagues remained for decades. Among them was Lawrence Martin, a young, but reputable geographer, that joined the faculty in 1906, but left academe to serve as a geographer in the U.S. Army after the outbreak of World War I. Martin concentrated primarily on geomorphology, continuing research in glaciology with Ralph S. Tarr, and editing the elder scientist's textbook on physical geography in 1914 (Williams 1956). Like Salisbury and Fenneman before him, here was a case of another luminary geographer leaving Madison in the prime of his career; however, before Martin's departure, a professor that had been on the faculty for nearly a decade remained to take charge of geography.

Having developed some expertise in the subject while studying under Ralph S. Tarr at Cornell, Ray H. Whitbeck arrived in 1908 to teach physical geography. Shortly after his arrival, he began to teach a course in economic geography, and for the first time both physical and human elements of the discipline were united under the same roof. Whitbeck was devoted to geographic education and accommodating the needs of public school teachers. He led the geographical side of the department until the 1920s, authoring a number of textbooks, and gaining some status in the discipline, actually becoming the President of the Association of American Geographer in 1925. Although a benevolent and kind gentleman, he did not attack geographical studies with any great rigor, and his influence on graduate students waned considerably. Glenn Trewartha (1978) commented

> My first content course with Mr. Whitbeck was taken when I was a graduate student. While interesting, it was still something of a disappointment, largely because it contained so much that seemed to me dubiously geographic. I was too complacent and was not stretched intellectually. In general, graduate students found Whitbeck less stimulating than did the less critical undergraduates and the teachers (p. 17).

By 1929, geography at the University of Wisconsin had gained autonomy from geology. Trewartha (1979) wrote, "Rumor had it that the process of separation was instigated when [Vernor] Finch was invited to join the geography faculty at the University of Chicago and, as bait to hold him at Wisconsin, there was proposed a promotion in rank and a separate Department of Geography was made" (p. 19). The geographers had already developed solid reputations; Finch, who was the institution's first geography Ph.D., contributed considerably to the geography of agriculture and authored several textbooks. By the end of the Second World War, thirty-four geography dissertations had been completed; only Chicago and Clark produced more doctorates during that time-period than Wisconsin.

Quantity, however, did not infer quality; between the World Wars, graduate instruction continued to be marked by mediocrity. Trewartha (1978) recalled:

> By present standards, the quality of the graduate training in geography at Wisconsin as of the early 1920's left something to be desired. Largely lacking were advanced courses specially tailored for graduate students. Only modest attention was given to methods and techniques of geographic research. Both Whitbeck and Finch offered graduate seminars, but only in Finch's seminar was the student obliged to prepare and read an original paper. The requirements for advanced degrees were vague and largely uncodified; as I recall, no preliminary examinations were required for the doctorate. . . . I must admit that in my early years at Wisconsin, the emphasis upon scholarly work by the three-man geography staff was not overly strong. It was Finch, mainly, who stimulated research efforts. . . . (p. 27).

Finch was one of the very few geographers of an earlier generation that completed graduate training in geography, rather than following the more common circuitous route by studying another discipline and then gaining entry into the field (Hartshorne 1961). He served as the department's chairman from its inception in 1928 until the end of World War II, and he facilitated a climate of academic integrity, scholarship, respect, and service. Hartshorne (1961) reflected shortly after his death:

> [I]n his relations with students, as well as staff members and other colleagues, emphasis on scholarship was always combined with great personal consideration, sympathy, and genuine kindness. The fruit of these labors is to be observed in the respect and affection in which he was held by his former students and in the positions of distinction which many of them now occupy in the faculties of many American universities (p. 340).

Finch collaborated with his former student, Glenn Trewartha, in the development of a textbook that revolutionized the teaching of the discipline at the introductory level in colleges and universities. It was their *Elements of Geography*, initially published in 1936, which would later undergo four additional revisions over the course of three decades that established the content and framework of physical geography instruction in the United States, based largely upon the outline from Carl Sauer's syllabus at Berkeley (DeVivo 1988; Leighly 1976).

In any event, one might wonder what would have happened had Finch gone on to Chicago in the 1920s. Ray Whitbeck was dedicated and popular among schoolteachers, and he was recognized by the geographical community as a scholar, for the Association of American Geographers required both evidence of scholarship and a nomination from a member inside the organization for membership, and Whitbeck achieved the highest honor the association could award by being elected its president. By the same token, his scholarship reflected the simplistic tone of an earlier era marked by speculations concerning environmental influences (or controls) on human activities, and his lack of graduate credentials (for he held only the B.S. degree from Cornell), in addition to an unwillingness to apply the necessary rigor, which characterized graduate instruction at Chicago or Berkeley, provide evidence that he was marked by a laissez faire leadership style. Finch, on the other hand, had applied geographical principles to resolving real world issues. "[I]t was Finch on whom the graduate group mainly depended for instruction, direction, and inspiration" (Trewartha 1978, p. 27), and it appears that it was Finch, characterized by some transformational leadership attributes, that set the ball in motion for Wisconsin to gradually develop as a reputable center for graduate studies in geography.

That is not to say that Finch was a transformational leader, for though he generated scholarship in regional geography and field studies that was rather noteworthy, he determined that the amount of time needed to conduct this type of research was not the most practical way in which to practice professional geography; thus, the volume of his research and publication over a rather long career was not especially striking. One geographer remarked that "Finch owed his good standing among geographers more to his sterling character, critical ability and sound professional judgment than to productive scholarship" (Trewartha 1978, p. 30). Moreover, though he illustrated his lectures in undergraduate classes with numerous slides, his monotonous delivery left many students uninterested, though more serious students found his classes to be full of substance. Still, his reputation as a dedicated chair was known, and he assumed these administrative duties for eighteen years with great resolve, realizing that the responsibility demanded much time; but by serving in this position, he acknowledged that subordinate members of the department would have more time for their own geographical research.

Finch did not appear to exude charisma, for his somber manner and dark attire kept many at a distance; though it must be noted that he was kind, but not an affectionate individual that generated warmth in relationships with colleagues or students. In fact, in writing letters of recommendation, when the norm was to exclude or minimize shortcomings, Finch would remark about an individual's strengths and weaknesses equally. This candor, a deviation from the norm, disadvantaged solid students from Wisconsin, for they often failed to get good positions they might have otherwise deserved.

Upon reflection, it appears that Finch was quite a paradox of sorts in terms of leadership. In comparison with his predecessor, Whitbeck, he was marked by some transformational behaviors because he elevated the graduate students to a higher standard. He demonstrated his scholarship early in his career, and received a number of honors: an honorary doctorate at his alma mater, Kalamazoo College, the Helen Culver Gold Medal of the Chicago Geographical Society, and election to President of the Association of American Geographers in 1938. It also would appear that he was at least partly responsible for the honorary doctorate that was awarded in 1937 by the University of Wisconsin to Isaiah Bowman, then president of Johns Hopkins University. Yet he was not charismatic by any means, and his inability (or unwillingness) to engage in scholarship while he acted as chair, seemingly failed to generate a high level of intellectual stimulus among students. He certainly did not invoke a level of inspiration demanded of transformational leaders.

On the other hand, he was considered a "leader generous of disposition and understanding of scholarship" (Martin 1994, p. 489). He did attempt to facilitate the success of his subordinate colleagues, by making recommendations for external grants, fellowships, and other opportunities; however, the matter of refusing to acknowledge the unfair impact his letters of recommendation generated is disconcerting. Given all these factors, it appears that Finch was marked with a combination of leadership behaviors ranging from transformational on the one hand to transactional and passive-avoidant on the other.[1]

Regardless, with his health failing, Finch relinquished control of the chairmanship to Glenn Trewartha after the end of the Second World War. In addition to other fields, such as climatology and economic geography, regional studies had already been firmly established by the Wisconsin faculty, and the department had achieved moderate growth as one of the few graduate programs in the country. Loyal Durand had joined the staff in the 1930s, along with J. Russell Whitaker, who was extremely popular in geographic education; however, the latter departed at the end of the decade to take a position at the George Peabody College for Teachers in Nashville.

Whitaker, with six years of teaching experience at the teacher's college in Marquette, Michigan, had been awarded the doctorate at Chicago and hoped to find in the department at Wisconsin "intellectual stimulus and support from the department colleagues; and to be particularly challenged by graduate students" (Martin 2003, p. 224). Instead, he found that developing a sense of comradeship among faculty and students was not considered a priority, and though he collaborated with his colleagues in publications devoted to geographic education, he longed for the opportunities to cultivate personal relationships as he had experienced in northern Michigan. He commented,

> I missed the more intimate relations with students that had been mine at Marquette. I sensed that I was unwilling to become the narrow specialist expected of me by my colleagues, and I now suspect that I lacked the independence and hardihood of spirit to ignore the climate of opinion in the Department (Martin 2003, p. 225).

Morale was lacking in the Wisconsin program; thus, when an opportunity came to teach the low-income students at the teacher's college in Tennessee, where he could also make an attempt to ameliorate racial disharmony on the social landscape, the die appeared cast and he moved his family to Nashville where he would remain for the duration of his career.

Richard Hartshorne, who had been on the faculty at the University of Minnesota for a number of years (after completing the doctorate at Chicago), was invited to join the Wisconsin department as Whitaker's replacement. Hartshorne had already developed a reputation for his work in political geography as well as the discipline's history and philosophical foundations in his 1939 masterful work, *The Nature of Geography*.[2] Shortly after Hartshorne's arrival, Durand left Madison to take a position at the University of Tennessee.[3]

The war interrupted graduate studies at Wisconsin, just as it did elsewhere, for not only did most prospective students join the war effort, so did many faculty; here, the role of geographers was significant, especially in intelligence analysis and operations; the Office of Strategic Services (OSS, the forerunner to the CIA) employed 129 geographers. In contrast, membership in the Association of American Geographers totaled 200 (Barnes 2006). Most of the geographers were devoted to analysis in Washington, DC; however, the discipline also was represented on the operations side.[4]

In any event, Trewartha remained at Madison for the duration of the war. He had been on the faculty since 1922 and had developed an eclectic body of work, for his research included studies in climatology, East Asia, demography, and settlement geography. Though he remained at Wisconsin for his studies as well as his teaching, he looked to Carl Sauer for a short time, as a philosophical mentor, and he embraced the regional concept, which was passed on to students regardless of whether the course of study was in the physical or human side of geography. Though not enamored with the idea of taking over as an administrator in charge of the department, Trewartha believed that it was the duty of the faculty to periodically share in the responsibility of leading the department as chair, and he set an important precedent by serving in that capacity for only five years, establishing the practice of short-term chairmanships from then on (Hartshorne and Borchert 1988).[5]

After the war, Arthur Robinson joined the faculty in Madison. Robinson had received the master's degree from Wisconsin, and he had also distinguished himself with cartographic work while attached to the OSS as a subordinate of Hartshorne, for which he was awarded the U.S. Army Legion of Merit, one of the highest non-combat decorations awarded for meritorious service by the armed forces. He recalled in an interview nearly forty years after the war how fortuitous circumstances led to him serving in the OSS; Richard Hartshorne had been called to Washington to serve as chief of the geography division of the OSS in September 1941. Interestingly, while traveling en route to his destination, he stopped in Columbus, Ohio, and discussed with the chairman of the department at Ohio State his need for a cartographer in this new government agency (Dow 1984).

> Dick had to move his family to Washington. Mrs. Hartshorne, I'm not sure why, took the train to Washington and Dick took their two daughters and drove to Washington in order to get the car to there. Well, it was summer time and hot and they stopped in Columbus. Dick and his daughters stayed overnight with the Peatties and let the kids get cleaned up and so on. During their visit, Dick asked Peattie did he know anybody in geography who knew how to make maps? He said: "Oh sure I do, he's been working for me." Dick talked to me, but about a month later in October or thereabouts I got a call from Washington. Would I come down and work with him in the budding Geography Division of the COI, which was the forerunner of the OSS?

Robinson was in the midst of completing the doctorate at Ohio State University under Guy-Harold Smith when he arrived on the Madison campus in January 1946 (Morrison 2008). Quickly gaining prominence as the dean of American cartography, he attracted a strong following among many students interested in the art and science of mapmaking.[6] Able to secure a large number of National Defense Education Act fellowships for advancing the study of cartography, he brought in substantial funding for students as well as the department for cartographic laboratory equipment and space. "Robbie" was asked about his role as "the guiding light" in bringing about changes in modern cartography. "No, no, no! I'm just one. There are lots of 'others" (Dow 1984).[7]

Following Trewartha's term as chair, the duty rotated to Hartshorne, then Robinson, and then to Clark to lead the department, and by all accounts, they did so rather ably, for a 1957 national study concluded that the University of Wisconsin was ranked as the top geography department in the country, followed by Chicago and Berkeley. Andrew Clark and Edwin Hammond, both holding Ph.D.s from Berkeley were among those that joined the faculty in the early 1950s, and each had served in the OSS during the War; there were also others.

Clark founded the geography department at Rutgers after the war before arriving in Madison as a full professor to continue his career in historical geography. Having studied at Toronto under Griffith Taylor and later under Sauer, he was regarded as a brilliant scholar in the field, and he attracted a large number of students anxious to test themselves before a demanding taskmaster. Of the nineteen doctoral dissertations he supervised, eleven were published as scholarly books, which remain as seminal pieces of historical geographic research.

Clarence Olmstead also saw service in the OSS during the war, and he joined the department in 1946 and received the Ph.D. from Michigan in 1951. He published much in the geography of agriculture, and by the time he retired after thirty-five years of service, he had served on the dissertation committees of seventy-five doctoral students. He was known for spending large amounts of time in advising students, and he also developed a reputation for seeking to extend justice impartially to all members of society.

Kirk Stone, also a former OSS analyst, John Alexander, an economic geographer with a solid reputation for teaching and research, and Edwin Hammond joined the faculty in the early 1950s. Hammond had recently completed his dissertation in geomorphology, and for the first time since Armin Lobeck's departure in the 1920s, the study of landforms was reintroduced to the geography curriculum at Wisconsin. During the war, Hammond had also served in the OSS as a geographer engaged in preparing intelligence reports before being commissioned as an officer in the U.S. Navy and seeing service in the Pacific theater of operations. The son of a professor at the University of Missouri, he was not unfamiliar with academe. He made some significant contributions to the study of landforms, with particular reference to terrain analysis; for example, his landform map for the 1970 *National Atlas of the United States* continues to be recognized as a valuable representation of the country's topography. Yet, the volume of his scholarship was limited, and it appears he did not generate a large following among students.[8]

In essence, the climate at Wisconsin changed significantly after the Second World War. With the loss of Whitaker before the war, a personable faculty member with solid leadership potential was gone; however, Durand's move to Tennessee appears not to have made a negative impact, nor did Finch's resignation from the chairmanship and subsequent retirement, as his shortcomings in leadership have already been noted. Trewartha appears to have been ready for the change, and he set the stage by limiting his role in the formal leadership position to a relatively brief period. At first glance, it would appear that with the addition of three stellar geographers in Hartshorne, Clark, and Robinson, new perspectives pertaining to developing relationships with one another and building a graduate program for students facilitated success, as the relatively new department was considered the best one in the country by the end of the 1950s. Along with substantive expertise in geography, one might think that the trio likely acquired transformational leadership skills from their mentors at Chicago, Berkeley, and Ohio State, as well as in the OSS, which served them, their colleagues, and their students in good stead; however, Karl Butzer's account of his early experience on the Wisconsin faculty tends to contradict this perspective, as he compares the environment there with his graduate student days at the University of Bonn, studying under the notable German geographer, Carl Troll.

> [T]he Department of Geography at the time was smug and authoritarian. . . . Most of the senior faculty had worked for the intelligence services during the war and had known each other well for at least fifteen years. They formed a good old boys' club, with a strong military flavor and redolent of rank. The chairman, Andrew Clark, was a Canadian. . . . in Bonn there had been an explicit hierarchy that comfortably structured formal relationships . . . in the institute or . . . at Troll's home, with the participation of a number of his nine grown children. His authority was based more on the affection of students and staff than on the institution. Not once did I see him act overbearing or wield his authority to patronize a subordinate; his critical mode was paternal and constructive, never ad hominem. The Madison department worked the other way around, with a fictive egalitarianism and all decisions made in private by only three to five power brokers, despite obligatory weekly faculty meetings that ran all of each Tuesday afternoon. Subordinate full professors either remained inconspicuous or were sarcastically put down. The graduate students were nervous and uptight, and displayed no spontaneous adulation for faculty (Gould and Pitts 2002, p. 70).

Butzer's commentary appears to be by all means legitimate. Forrest Pitts also refers to a number of students from the Madison campus that sought to shift their focus from Wisconsin to Ann Arbor for the pursuit of graduate studies (Gould and Pitts 2002).[9] In the final analysis, it appears that the Madison department in the 1950s was dominated by selected members of the faculty, such as Clark, Hartshorne, and Robinson, who had seen service in the war and had already established for themselves solid reputations in research and publication. The departmental leadership was more transactional than transformational; yet reputation in scholarship facilitated the program's standing, and numbers of graduate students arrived ready to study. Apparently, several did not look upon their experiences fondly, and when they went to serve as faculty members in graduate programs elsewhere, a marked absence of transformational leadership attributes characterized their behaviors as well.[10]

NOTES

1. In 1986, I spoke with Edwin Hammond at the University of Tennessee, who had arrived at Madison to begin graduate studies in 1939. Apparently, some sort of graduate student exchange program took place between the campuses at Madison and Berkeley in the late 1930s and early 1940s. Some stellar graduate students from one university would be invited to study at the other for an academic year, and the intention was for them to return to their home institution to complete the Ph.D. Hammond was asked to take part in this program, and he accepted while on the verge of completing his master's degree at Wisconsin. When it came time for him to return to Madison, he chose to remain at Berkeley; even the offer of the prestigious Whitbeck fellowship could not entice him, for the Berkeley department, led by the challenging taskmaster, Carl Sauer, who was characterized by a markedly high level of transformational leadership qualities, was far more appealing than the Wisconsin department led by Finch.

2. Richard Hartshorne appears to have developed an unfair reputation for not tolerating opposing viewpoints; yet, time and again I have heard from former students and colleagues that this was simply not true. David Ward related to me that he was not "adverse to contrarian students" and "never censorious about students' other views." Brian Berry (2006) recounted how he delivered an invited lecture at the University of Wisconsin during his early years as an assistant professor at the University of Chicago and was met with criticism by a number of faculty; however, it was Richard Hartshorne that showed support for Berry and rebutted his critics, despite their philosophical differences.

3. I was acquainted with one of his doctoral students from the 1950s, the late Sydney Jumper, and was surprised to learn through the course of conversation that although Durand had engaged in scholarship and was particularly known for the geography of dairying, as well as writing textbooks, he appeared not to be a particularly engaging teacher or mentor.

4. In 2002, at the annual meeting of the California Geographical Society in Lone Pine, Bill Denevan and I talked at length about geographers of the past. Bill had retired from the University of Wisconsin; a notable scholar in his own right, he had been a student of Carl Sauer and Jim Parsons. He told me about an account of a wartime experience related to him by Andrew Clark, also a student of Sauer and also a geographer that spent the bulk of his career at the University of Wisconsin. Apparently, Clark was attached to the OSS in Asia, and on a particular operation, he happened to be seated next to the future Communist Chinese Premier, Zhou Enlai, on board an air-

craft in frigid temperatures, "crossing the hump" over Burma. My conversation with David Ward a dozen years later corroborated this fact about Clark's wartime service; in fact, Clark was not an analyst, but an OSS operative commissioned in 1944 to determine the nature of Chinese Communists and to see if the most European of Chinese Communists could be swayed. His conclusions were that his traveling companion was very orthodox in his beliefs, and history has revealed what Clark noted. Interestingly, Clark's obituary makes no mention of his wartime service (Ward 1977).

5. David Ward indicated to me that context is very important in examining the history of the department at Madison, and here is a good example, for at about the same time Trewartha was endorsing the notion of the rotating department chair, the university was moving in that same direction.

6. Leon Yacher (1982) has presented a strong case for Hungarian-born Erwin Raisz as the father of American cartography. Among other things, Raisz developed the first cartography textbook and developed a large number of hand-drawn maps, many of which are still in use today. Unfortunately, Raisz struggled with the English language, and although his work was truly captivating—so much so that the janitorial staff on occasion refused to erase the block diagrams and raised relief maps he drew on chalkboards—in his position at Harvard, he did not have a forum in which he could attract graduate students. At Wisconsin, Robinson had that opportunity.

7. Other centers for the instruction of cartography were located at UCLA under Norman Thrower, Washington under John Sherman, and Kansas under George Jenks.

8. I was acquainted with Hammond late in his career and had discussions with some academic geographers that were in his classes while he was on the faculty at Madison. Those disinclined to study physical geography, such as Sam Hilliard, found his teaching uninteresting, whereas those interested in the subject, such as Bob Richardson, were attentive. Although he was not the most captivating speaker, and his meticulously prepared lectures did little to spark interest among most students, they were content-rich, and he devoted much time and effort toward developing innovative approaches in the study of physical geography. Regardless, Hammond told me that he was unhappy with the prospects for promotion at Madison—he remained an associate professor after twelve years there—and he departed to take a position on the faculty at Syracuse in the early 1960s.

9. I recall discussions with former graduate students at Wisconsin during this period that indicated they resented the authoritarian manner in which Clark had not only administered their graduate studies, but also their early careers. For example, Leonard Brinkman (Ph.D. Wisconsin 1964) related to me how he had not really been given a choice as to where he would go to begin his academic career as a non-tenured faculty member; Clark wanted him to go to the University of Melbourne, and there he went.

10. It might be unfair, however, to view the oligarchs at Madison as a malevolent cabal, for the behavioral attributes of these senior faculty were probably more the norm than the exception in American academe during the 1950s, '60s, and '70s, and seemingly typical of nearly all University of Wisconsin academic departments. Senior faculty shaped each department's destiny. That being said, Butzer's dislike for Clark and Robinson, very close friends, was obvious, albeit perhaps justified. Moreover, just because the "fictive egalitarianism," which typified Madison may have been the norm in American academe, it did not mean it was acceptable. Yet, things change. By the end of the 1960s, the department was relatively young, and according to David Ward, "There was an overtly transparent style of doing things."

FIVE

Lackluster Leadership at Michigan

During the first two decades of the twentieth century, geography built momentum in Midwestern public universities as a respectable discipline, as geographers were collected in departments, usually binomial in character, and often with geology; but their focus represented a marked departure from earth science. At the University of Michigan, William Herbert Hobbs, known for his studies in glaciology, became head of the geology department in 1906. He had been on the geology faculty at the University of Wisconsin when T.C. Chamberlin was president of the institution before departing to the University of Chicago. Several years later, after promotion to the rank of full professor, and after the death of the geology department head at the University of Michigan, Hobbs was asked to replace him.[1] The move to Ann Arbor was made, and as Hobbs actively participated in meetings of the Association of American Geographers, he further demonstrated commitment to the discipline by bringing geography into the university curriculum.

In 1915, Hobbs hired one of Rollin Salisbury's students from the University of Chicago, Carl Sauer, to serve on the faculty in a new binomial geology and geography department. Sauer had taught for a year at the normal school in Salem, Massachusetts, before completing his doctoral dissertation; after accepting the appointment at Michigan, he apparently led the geography side of the department, although the distinguished Hobbs continued to serve as chair. In retrospect, it is obvious that his independent spirit and leadership were apparent at Michigan, for though his tenure there lasted only until 1923, he gained a full professorship and the following of devoted students (Parsons 1976). Some followed him to Berkeley, and they became luminaries in their own right.

Fred Kniffen, who studied geology at Michigan during Sauer's tenure there, made some comments when asked about his experience:

> In the first place, Carl Sauer was in Ann Arbor at that time. While I never had a course with him, we had long talks. He had an office across from mine when I was an assistant in mineralogy. We talked about a great variety of things. I don't recall that he ever mentioned the name geography, but he talked about the things that I was interested in (Dow 1976).[2]

Sauer captivated students in a number of ways, one of which was through field studies that were devoted to assessing land use in Michigan, especially in areas that had been heavily deforested. Sauer engaged in numerous discussions with land economists, foresters, and state conservation officials, and he developed quite a bit of knowledge about the destructive exploitation, which had occurred at the hands of lumbermen, finding that the landscape was often unsuitable for agriculture. He directed students in land use and mapping surveys around the state to assess then-current conditions, which proved to be not only especially captivating, but also useful for the state to conduct sensible conservation planning (Leighly 1976; Martin 2005). This type of regional analysis was a significant departure from the sort of geography that was influenced by the environmental determinism, which remained a major focus in geographical scholarship at the time, as well as geomorphology, and Sauer's leadership attracted a number of students to follow his path. Before leaving for California, he asked one of his graduate students, John Leighly, to join him, who also became one of the discipline's leading luminaries. Incidentally, Leighly, who had served in France during the First World War, may not have had any other options if he chose to continue graduate studies, for he was dismissed from his assistantship in geology. He incurred the wrath of Hobbs for placing his feet on his desk while talking to a student and taking too many courses on the geography side of the department. Thus, Leighly left the University of Michigan and went on to Berkeley with Sauer, completing his Ph.D. in 1927.

Along with the departure of Sauer went Darrell H. Davis, a new Ph.D. from Michigan, who left for the University of Minnesota. The need for a replacement was obvious, and in came Preston James, a new Ph.D. from Clark University. James had studied under Wallace Atwood at Harvard and found him to be a particularly inspiring professor. He later completed a master's degree while studying climatology under the eminent Robert DeC. Ward, and though James contemplated remaining at Harvard for the doctorate, when Atwood accepted the presidency at Clark with the intention of building a geographical institute, he again studied under Atwood and completed the degree in two years.[3] When the opening at Michigan occurred, Ellen Churchill Semple, a member of his dissertation committee, wrote an excellent letter of recommendation to Hobbs on James's behalf: "He has a pleasant personality, good presence, a clear ringing voice in lecturing and great enthusiasm for geography. He is one of the boys whom I am fond of" (Martin 1988, p. 166).

Now separate from the geology department, which continued to be led by Hobbs, for the Board of Regents approved an independent geography department in 1923, a chair was sought for the program. Appeals were made to Charles Colby and Wellington Jones, both on the faculty at Chicago; but they refused the offer. Consequently, K.C. McMurry, at the age of twenty-eight, took the helm of the newly created department made up of young geographers, all under the age of thirty. This was a significant departure from the authoritarian manner in which the department was led by Hobbs, for McMurry did not appear to dominate the small group that made up this new program.[4] James commented:

> McMurry was a young fellow who was the Head of the Department. Then there was Bob Hall, Stanley Dodge and myself—all young fellows. We had a great time building a new department, as it were (Dow 1984). There was no autocrat over us. Nobody standing in authority, but all of us could have an opinion. We expressed ourselves freely . . . the faculty and the graduate students would get together at frequent intervals and exchange ideas about geography. . . . We met once a month at someone's house (Martin 1988, p. 166).

McMurry was known for his fieldwork in cartographic analysis and mapping with special reference to employing aerial photographs. Having completed his doctorate at Chicago in 1922, his work was considered highly innovative, for he integrated analysis of aerial photographs of Isle Royale with field observations to determine the land cover the aerial images represented. This made it possible to reduce the labor intensive efforts for later surveys of land use (Martin 2005).

In any event, by the time the Second World War ended, the Michigan program ranked fourth in Ph.D. production among American universities, with twenty-three awarded by 1946 (Rugg 1981). The faculty was productive in scholarship as well as fostering geographic education; for example, James's 1935 introductory college textbook, *An Outline of Geography*, was included among the forty-eight publications he authored while he was at Michigan. Here, as in the case of Finch and Trewartha's textbook, Carl Sauer willingly provided notes that proved to be extremely helpful to James (Martin 1988).[5]

James also showed effectiveness as a mentor. Realizing the difficulty faced by young Kirk Stone, a graduate student struggling to support his family while a graduate student attempting to complete the doctorate, James recommended him for a position at the University of Toledo. More than a decade later, he completed his dissertation at Michigan and subsequently accepted an appointment at Wisconsin following his World War II service in the OSS (Dobson et al. 1999). Stone's record in academic geography was one of distinction.

James, an officer in the Army Reserve, also joined the OSS at the request of its director, William Donovan, as the country contemplated war in the autumn of 1941.[6] There he worked in the Research and Analysis branch along with a number of other colleagues in the discipline, and his influence was pervasive. In fact, he recommended Hartshorne for the position of Chief of the Geography Division (Martin 1988). After the war, James considered returning to academe in Ann Arbor, but he was offered a very attractive position on the faculty at Syracuse University by the chair, George Cressey, which he accepted.

Donald Brand, a 1933 Ph.D. from Berkeley, arrived in Ann Arbor as a full professor in 1947, apparently taking on the responsibility of James's courses and research in Latin America. He attracted to the program an Austrian immigrant that had fled from the Nazis, who had also engaged in work for the OSS: George Hoffman. Hoffman completed his doctorate in 1950, and then joined Brand when he was invited to chair a new geography department at the University of Texas in 1949 (Knapp 1998).

George Kish, an immigrant from Hungary who already held advanced degrees from the Sorbonne and University of Budapest, came to the University of Michigan as a research assistant, gaining the Ph.D. and becoming a member of the faculty in the middle of the War. Kish remained at Ann Arbor throughout his career guiding a number of graduate students in the course of their studies, as did other members of the faculty, such as K.C. McMurry and Stanley Dodge.

Still, no leader in the department, either formal or informal, could capture the hearts and minds of graduate students and infuse them with a passion for geography as did Carl Sauer, and his departure in the early 1920s would be felt for generations. Preston James certainly created a strong presence on the Ann Arbor campus; but even he, who gained luminary status in the discipline and assisted a number of students in their dissertations devoted to regional studies in the 1930s, could not match the charisma, influence, and transformational leadership of Sauer. No one could.

In contrast, despite his many accomplishments, the evidence appears to indicate that K.C. McMurry, thrust into the role of chair early in his career, was a rather passive leader more inclined to embrace a laissez faire manner of heading the Michigan department than anything else. Granted, his landmark studies in land use assessment, air photo interpretation, and recreation are noted, as well as his participation in the greater academic community, as exemplified by his work with the Social Sciences Research Council (Lasswell 2005). Yet the Michigan department under his leadership might have had a better reputation than it truly deserved, especially with the loss of James after the War. Many doctorates were awarded in the 1940s and 1950s of variable quality. Though a handful of students went on to gain solid reputations in the discipline (e.g., Clyde Kohn 1940, Kirk Stone 1949, George Hoffman 1950, Clarence Olmstead 1951, Forrest Pitts 1955, and Peirce Lewis 1958), most did not. Dissertations appeared to follow no specific theme or particular focus, as they had in the 1930s, which demonstrated the eclectic nature of the discipline, as well as the department's faculty, and they varied in length considerably. Olmstead's requirement was well over 500 pages, and Peirce Lewis, who later achieved luminary status in the discipline, barely wrote 150. Certainly, dissertation length was not necessarily reflective of quality; but it begs to question if the several tomes generated at Michigan of less than 200 pages would have been approved by Clark at Wisconsin or Sauer at Berkeley, who had by the 1950s established international reputations for producing leaders in the field well-grounded in geographical scholarship.

Essentially, it appears that the Michigan program was characterized by a lackluster leadership in which McMurry served largely as the administrative chair, seeking consensus from departmental members in major decision making efforts. This model appeared to work well during the pre-World War II period when James's dynamic personality sparked some intellectual stimulus from a position as an informal leader, but after the war, this type of leadership did not seem to be effective. With the exception of Brand's brief tenure, there was no time in which a transformational leader emerged, either formal or informal. Consequently, the climate of the department, marked by the acceptance of laissez faire leadership attributes, which were established early in its history, would present far reaching ramifications for the future of geography at Michigan.

NOTES

1. Hobbs served as the department head for twenty-eight years.

2. When I talked with Kniffen in Louisiana, now nearly thirty years ago, he commented that the geology students and geography students both conducted field studies. Though the geologists were deeply involved in examining the natural landscape, there was some question as to what the geographers were doing. Kniffen commented with a chuckle that "Driving around, looking at cornfields" appeared to be what geographers' fieldwork was all about. Of course, the esoteric nature of the discipline must have lent much to this view, for it likely seemed unreasonable to a student of a rather narrowly defined discipline then, as it often does today, that a practitioner of a field as eclectic as geography can integrate elements drawn from a number of disciplines, examine a portion of the earth, and deliver cogent analysis.

3. At Harvard, Atwood was known for awarding overly generous grades. There is little evidence to indicate that his lax standards changed after his move to Clark, where he established a new geography program that produced more Ph.D.s than any other before the end of World War II. Although it happened elsewhere, the two-year Ph.D. achieved by James, who had just reached the age of twenty-four, is a case in point.

4. Still, as things often go in academe, though officially outside the department, the influence exerted by Hobbs upon the newly autonomous department remained substantial (Hall 1953).

5. Although Finch and Trewartha introduced the Koppen scheme of climatic classification in introductory physical geography in 1936, and James did the same a year earlier in his *Outline of Geography*, Sauer really deserves much of the credit. It was from

the notes Sauer had been using since the early 1920s, which discussed the Koppen scheme, that Finch, Trewartha, and James got the idea.

6. James discussed how he became involved in military intelligence in an interview with Dow (1970): "During the First World War I went to Camp Lee, Virginia to the Infantry Officers School. I had been in the Harvard ROTC. At the end of the war (1918) I had almost completed the Officer Training Program and like many other people I immediately left and went home. But Lawrence Martin, who was at Clark, had been an intelligence officer and was still a Lt. Col. in intelligence. He said: 'You're crazy not to get into this field and maintain a reserve commission.' I, therefore, took out a 2nd Lieutenant's commission in military intelligence and I maintained this active all during the period between the wars."

SIX

From Tyranny to Transformational Leadership at Minnesota

When Darrell Davis left Ann Arbor for the Twin Cities, he was forty-four years of age, holding a new Ph.D., which was awarded the year of Sauer's departure for Berkeley. This was the first geography doctorate to be awarded by Michigan in the twentieth century, a marked contrast to the dozen or so that had been awarded by the more prestigious University of Chicago where master's theses, which often could have been dissertations for the Ph.D. at other institutions, were far more numerous than doctoral dissertations; but still, Davis was one of the relatively few holders of the geography doctorate in the United States. The following year, Richard Hartshorne, twenty years younger than Davis and holding a Ph.D. from Chicago, arrived to begin his academic career as an instructor at the University of Minnesota. By the end of the decade, Ralph Brown, holding the doctorate from Wisconsin, and Samuel Dicken, who was completing his dissertation under Sauer at Berkeley, joined the small department. With the addition of these three young geographers, each of them luminaries by the ends of their careers, the Minnesota program had the potential to be one of the best graduate programs in the history of geography before the onset of the Second World War; but this was not to be.

Hartshorne had studied mathematics at Princeton as an undergraduate, and he sought to pursue graduate work in geography under Ellsworth Huntington at Yale, but the Ivy League institution had essentially eliminated its program. Consequently, he went to Chicago where he studied under giants in the field; included among them were: Harlan Barrows, Charles Colby, Wellington Jones, Ellen Churchill Semple, J. Paul Goode, Henry C. Cowles, and Robert Platt. Colby arranged for him to be hired by the city of Chicago to conduct a study of local water transportation, which became the subject of his dissertation, "The Lake Traffic of Chicago," and in the meantime, Barrows recommended him for the faculty appointment at the University of Minnesota (Dow 1972; Martin 1994).

When Hartshorne arrived in Minnesota, he found a marked contrast to the gregarious stimulating environment that he had experienced in Chicago. Though Darrell Davis, the chair, sought to develop a stellar program, he was inclined to keep to himself and work in solitude, making departmental decisions without discussing things with faculty at all. Thus, finding no intellectual stimulus (or camaraderie) in Minnesota, Hartshorne pursued involvement with the Association of American Geographers and participated in a number of meetings and field conferences. His research became highly regarded, as he published articles on location analysis, transportation costs in relation to iron mining and steel production, and urban geography. When he shifted his focus in the 1930s, he became one of the few students of political geography in the United States, while also conducting research in Europe (Martin 1994).

Discourse concerning the essence of the discipline had emerged in conversations at conferences, papers presented at annual meetings, and a number of publications, such as Sauer's (1925) "The Morphology of Landscape," Colby's (1936) "Changing Currents of Geographic Thought in America," and Leighly's (1937) "Some Comments on Contemporary Geography Method." Sauer's message created a considerable impact; but because it threatened the legitimacy of political geography, Hartshorne objected to his notion that "if you could not see it, then it was not geography, [and replied] 'It is time to challenge the great god west of the Sierras'" (Martin 1994, p. 484). Derwent Whittlesey, the editor of the *Annals of the Association of American Geographers,* invited Hartshorne to offer a philosophical commentary for publication in the flagship journal. Hartshorne submitted his work, and it was favorably reviewed by Robert Platt and Wellington Jones;[1] however, Hartshorne's chair at Minnesota, Darrell Davis, without giving reasons to back up his views, wrote to Whittlesey stating that his colleague's commentary should not be published (Martin 1994).

Whittlesey then appealed to Hartshorne to write a more expansive essay, and though he was captivated by the idea, he had other plans; Hartshorne intended to spend his sabbatical in Europe studying boundaries in the Danube valley. On board a ship that was preparing to go across the Atlantic, Whittlesey met him, and before the vessel departed from Boston, Whittlesey promised Hartshorne that his manuscript would be published in the *Annals*.

In a turn of events, soon after Hartshorne's arrival in Europe, the threat of war interrupted his work in political geography; so he decided to devote his sabbatical to conducting research in the history and philosophy of geography, a field that had been studied by a number of German scholars, but was not closely examined in North America. Hartshorne sought input from his German-speaking contemporaries, and he also spoke with Fred Kniffen, one of Sauer's early students, who he happened to run into at a conference in Germany. The end result was a manuscript of more than 400 pages, which was published in the *Annals of the Association of American Geographers* in 1939 as "The Nature of Geography" (Dow 1972; Martin 1994).

At about the time Hartshorne returned to Minnesota, J. Russell Whitaker left Madison for Nashville, leaving a position open at the University of Wisconsin. After heavy deliberation, Hartshorne decided to make the move to Wisconsin. It was a difficult decision. On the one hand, his colleagues, Ralph Brown and Sam Dicken had become lifelong friends, and his wife was the daughter of the chairman of the philosophy department. He had developed good relationships with a number of faculty at the institution. "But the headship of the geography department by Darrell H. Davis cast a pall. Davis could be insensitive and authoritarian" (Martin 1994, p. 489). Here, a tyrant as chair led to the departure of one of the most accomplished academic geographers of his generation.

With the departure of Hartshorne, Brown and Dicken lost a colleague and an ally. Both had conducted ample scholarship in their respective fields and taught a variety of courses, but it was Brown that had a far reaching influence through the relationships he developed with students as well as his scholarship. In reflecting on his mentor, Fred Lukermann recounted,

> I think Ralph Brown was unique in the sense of being a master teacher. I don't know how to explain this exactly, but the first course I took from Brown was at the junior level, I had had no geography before then. It is difficult for any student to march into what supposedly is the middle of the field. Yet, I accomplished it and Professor Brown probably made it possible. He had an enormous talent in terms of lecturing and discussion in classroom presentation, a very complete picture of what he wanted to do each lecture. Complete appreciation of the division between detail, individual events, range and variety of what he was speaking about, and also its general importance—the way it fitted into the general context of American history and so on. When I took the course I didn't know who Ralph Brown was (his background) but it soon became apparent. Not that he would tell you much about his own academic background or the literature that he had produced. The very interest he put into his lectures led you to read his material. I think it was from that point of seeing a scholar operate with the full status and image of a scholar (at least to the student in that class) that leads you into a discipline—if you are seeking to enter a higher education to do this sort of work for the rest of your life. Brown was a very quiet man, a very gentle man. He smiled teasingly at you at the end of classes, at the end of exams, at the end of conversation. But in between he gently poked you, pulled you, projected you along, and really he was an intellectual guide is the best way to put it. To the subject matter that

> course was one on the historical geography of North America and from that point I went on to the other subjects in the curriculum (Dow 1971).

Brown not only wrote a number of articles devoted to historical geography, but he also wrote books on the subject applying innovative approaches to the study of past landscapes, which also captivated students and led them to study historical geography. In fact, it was his (1948) *Historical Geography of the United States*, which remained in print until the early 1990s, that played a role in inspiring luminary Donald Meinig to change his career choice from diplomacy to historical geography (Meinig 1989). Moreover, he continues to be highly regarded and is commemorated by the Historical Geography Specialty Group of the Association of American Geographers through the *Ralph Brown Award*. Unfortunately, his premature death of an apparent suicide was a loss to the university community as well as the discipline. The *Geographical Review* (1948) reported an obituary:

> Ralph Hall Brown, professor of geography at the University of Minnesota, died in St. Paul on February 23, 1948, in his 51st year. As a leader in the civic affairs of the community of which he lived, as a respected and sought-after member of the university faculty of which he was a part, and as an instigator of scholarly inspiration in the world of professional geography at large, his influence had gained major breadth and stature. Among a host of individuals, as well as organizations and institutions, that have profited from his generous gifts of time and counsel, his loss seems very great (p. 506).

The loss of Dicken, who accepted the chairmanship at the University of Oregon the year before, also resulted in a significant change, as did the retirement of Darrell Davis within months after Brown's death. John Weaver, a Ph.D. who had served as a U.S. Naval Intelligence officer, joined the faculty immediately after the war, and Jan Broek, a Dutchman who had studied at Berkeley, but received his Ph.D. from Utrecht, took over as chair when Davis retired in 1948. A year later, John Borchert, who had served in the army in the European theater as a weather forecaster on a bomber base, joined the department after being awarded the doctorate at Wisconsin. These three individuals formed the crux of the department that within a very short period of time had lost the members that had built it from the ground up. Borchert commented in a much later interview, "When I got to Minnesota we were a very small staff, Broek, Weaver, and I; there were three of us really at that time. I was the third man, which meant I taught everything that nobody else wanted" (Dow 1975).

Broek had served on the faculties of both Berkeley and Utrecht before arriving in the Twin Cities, and in the Netherlands, he was considered "a teacher of high quality who could clearly explain complicated theories" (Ormeling 1975, p. 4); however, as chair and senior to his subordinates in the department, he appeared to be somewhat authoritarian in his leadership. John Adams, a student in the 1950s reflected on how a change in leadership took place.

> Borchert and Lukermann, along with some of the others were busy reacting to some of the heavy handedness that Broek had wheeled as Head (1948–56), when he apparently had insisted on advising all the Ph.D. students. Borchert had an offer to leave Minnesota and the whole fuss was settled when Borchert become Chairman, then Mather three years later as rotating chairmanships became the norm. I came into the department while Mather was Chairman (Dow 1983).

Borchert was known as an effective and just chair. By all accounts, he exhibited transformational leadership attributes. He was able to add new faculty to the department and develop a master's program oriented around applied geography in urban planning; twenty-two of his advisees went on to earn doctorates, and his obituary is marked with a number of favorable comments from former students. Among them, it was said:

> "He personified all that was meritorious in university teaching, student advising, research, and all that is good in a caring and loving professor."
> "He has been the most influential person in my life and the man I have most admired and tried to emulate."
> "Continuing interest in advisees throughout their careers."
> "Always positive and optimistic, but never critical, that was his great strength" (Adams 2007, p. 647).

In any event, Cotton Mather, who had taken the Ph.D. at Wisconsin after wartime service in the OSS, had served on the faculty at Georgia before joining the geography department at Minnesota in 1957. Like Borchert, he was known as a compassionate and caring leader among both colleagues and students.

> As department chair, Mather played an important role in gathering faculty and graduate students to propel Minnesota to a position as one of the top geography programs in the nation. His stature as a scholar, his genial personality, his thoughtfulness toward others at all times, and his remarkable ability to conduct meetings smoothly but with expedition were especially effective in developing a very good program. His sound judgment, his remarkable tact with members of the faculty, his unfailing and unfeigned courtesy, and his deep enjoyment of people all came forth in this challenging position.
> In a university filled with celebrities with supersized egos, Mather's distinguishing characteristics were his humanity and his modesty. He spoke firmly but deftly. He picked friendships, not fights. In a university culture that often relished adversarial proceedings, he often saw beyond two sides to new dimensions and fresh perspectives. He preferred irony and humor to scoring points (Karan 2003, p. 488).

Essentially, by the end of the 1950s, the department was in good hands with Mather as a chair; this represented a stark contrast to its leadership at the beginning of the decade while under the administration of Broek, and certainly throughout most of its history under the authoritarian reign of Davis. Under the leadership of Mather, the department was able to develop further as a hospitable environment for the promotion of scholarship and the exchange of ideas while providing an exemplary education for future practitioners of the discipline; this was not possible under despots like Davis and Broek.

In essence, early in its history, pearls had been cast before swine, for Davis had the essential ingredients of an extraordinary program in Hartshorne, Brown, and Dicken; but his very apparent jealousy, underhanded behavior, and absolute autocratic character worked to undermine the department's mission. It is difficult to classify this type of dictator; however, it appears that Davis exhibited a number of passive avoidant behaviors, marked particularly by his extreme sense of isolation; in a sense, he was a laissez faire leader that very well may have been afflicted with paranoia, for he also made decisions, not with the intention of supporting his subordinate colleagues, but because he carried envy toward them. He sought to elevate himself to a higher standing than them, not just by maintaining an inactive role in their careers, but by working against them. His letter to Whittlesey concerning Hartshorne's essay is a case in point.

The situation in the Twin Cities was noted by Thornthwaite (1961) in his assessment of the scholarly productivity of American geographers in 1939. He developed a rubric and commented on the wide gap in scholarship that characterized some departments, noting that the highest score of forty-five was achieved by Hartshorne and one of the lower scores was reflective of Davis: "Most conspicuous is the situation at Minnesota, where a man of grade 45 is supervised by one of grade 3" (p. 11). It is reasonable to presume that Davis's role as a department chair of questionable ability would have been discussed at professional conferences among members of the intellectual elite, and as Hartshorne's status became elevated within the discipline by producing highly regarded scholarship before he turned forty, Davis's credibility must have waned. By the end of his chairmanship, the department experienced 100 percent turnover. Perhaps if it was in a private institution or if it occurred in a different era, the department would have folded; however, with vast numbers of returning servicemen seeking to gain undergraduate degrees after the War, the need to provide them with undergraduate degrees worked in the department's favor, and it survived.

Subsequently, Broek came on board as the chair; but he lasted only a short time because of his autocratic behavior. The evidence is quite clear that Borchert[2] exhibited transformational leadership qualities, and these were the necessary ingredients to thrust the Minnesota program into distinction as a stellar geography department.

NOTES

1. On the one hand, the review of Hartshorne's work by his former professors at Chicago can be viewed with skepticism, for in academe today, this would be considered questionable at the very least. On the other hand, the fledgling discipline was represented by only a small number of doctoral programs where this kind of research was explored, and Chicago was clearly the leader. Also given that Whittlesey was an alumnus and former faculty member at Chicago himself, his acquaintanceship with Platt and Jones quite likely facilitated the review process.

2. Both Borchert and Mather were known to be personable characters able to talk with individuals from all walks of life. Their confidence in dialogue and relationship-building was observed by their students in field studies, whether in cities or rural areas. Arguably, these are traits essential for practicing transformational leadership.

SEVEN

The Struggle for Distinction at Ohio State

In 1924, the geography department at Ohio State University achieved autonomy. It was led by Charles Huntington, a member of the institution's faculty since 1909 who had been awarded the Ph.D. in industrial history from Cornell in 1915. As the department's roots were in the College of Commerce and Journalism, early courses took on an economic focus. Eugene Van Cleef, who would be awarded the Ph.D. from Clark in 1926, had joined the university as an economic geographer in 1921 and took charge of broadening the commercial focus of the curriculum. Fred Carlson, holding a Ph.D. from Cornell in soils joined the department in 1923, and after being on the geology faculty for four years, Roderick Peattie transferred from OSU's geology program to the department of economic and social geography.

Peattie had seen service in France during the First World War before completing the doctorate at Harvard in 1920. Interestingly, he expressed reservations about the transfer at first, responding to an appeal from Huntington: "I therefore, with considerable reluctance, decline your considerate offer." Yet, four days later, he changed his mind: "I am happy to report that as far as I am concerned final steps were taken towards the joining of the two departments when I telephoned you today to place my name on your staff list" (Smith and Taaffe n.d., p. 8).[1]

Because Peattie was awarded a one-year leave of absence in 1927, Guy-Harold Smith was brought in on a temporary appointment. He took a position at the University of Illinois for one year before returning to OSU as an assistant professor in 1929, now holding a Ph.D. from Wisconsin. Huntington, Van Cleef, Peattie, and Carlson all reached the rank of full professor by this time, and other members of the department included a number of instructors and teaching assistants that did not hold the doctorate.

By the end of the 1920s, the department was staffed by more faculty holding the Ph.D. than any other geography department in an American public university, and its members maintained a level of professional activity that matched, if not exceeded other programs; for example, in December 1929, OSU hosted both the National Council of Geography Teachers and the Association of American Geographers for their annual meeting, and the geography faculty continued to contribute much in geographical scholarship. Between 1926 and 1934, twenty M.A. degrees and the department's first Ph.D. were awarded, Huntington and Carlson coauthored a book, Van Cleef published forty-two articles, Peattie amassed six publications (including a textbook), and Smith contributed twenty-four works, including a number of maps. At their own expense, three members of the department participated in the International Geographical Congress in Paris in 1931; but salary decreases of nearly 20 percent prevented them from being able to take part in an international conference in 1934. Huntington then stepped down as chair, and Smith was promoted to associate professor and appointed to head the department, finding it a difficult struggle to balance the budgetary allocation with the needs of the department in efforts to achieve the institution's mission. He asked the administration for funds to raise the salaries of faculty, but was time and again denied.

Before the outbreak of the Second World War, Shannon McCune (Ph.D. Clark) and F. Webster McBryde (Ph.D. Berkeley) joined the department as instructors. By 1941, the department had awarded another twenty M.A. degrees and six Ph.D.s, and the faculty contributed a number of publications to the geographical community in the form of scholarly articles and books; yet with the onset of war, enrollment decreased and the departmental mission changed significantly. Huntington retired in 1943, at the age of seventy, and some members of the faculty took positions with the federal government to contribute to the war effort. For example, McCune was assigned to the Board of Economic Warfare, McBryde worked in the Military Intelligence Division of the War Department, and Peattie was stationed in South Africa with the Office of War Information. A number of geography graduate students also went on to work for the government in intelligence, such as Arthur Robinson and E. Willard Miller.[2] Still, even with the war going on, the department was able to award three doctorates.

When World War II ended, an influx of veterans pushed enrollment high in all courses. Across the country, Ph.D.s in geography were needed to serve in higher education, but they were in short supply and professors were tasked with carrying high burdens in teaching and supervising graduate students. John Randall, who had served as an instructor in the early days of the war, but had gone on to work with the Department of

State, was summoned back to Ohio, now as an assistant professor. Teaching loads remained excessive, and it adversely influenced scholarly research and professional activity; however, even in these circumstances, the OSU geography faculty continued to contribute an admirable amount to the geographical community. Peattie edited a number of books, Van Cleef authored several pieces, and Randall collaborated with Smith to complete an intelligence study of Iran. Moreover, in 1946, the National Council of Geography Teachers and the Association of American Geographers met again in Columbus.

In an unfortunate turn of events, Peattie, who was known for exemplary teaching, suffered a stroke. His lectures became "much less effective than previously when he was a fluent and interest-inciting teacher. It required a period of several years before he regained his teaching effectiveness. For several academic terms following his stroke, some members of the staff had to be ready to meet his classes when he was incapacitated or felt that he was not equal to teaching his courses" (Smith and Taaffe n.d., p. 28). Eventually, it appears that Peattie nearly fully recovered after several years, and in 1954 he returned to Africa for fieldwork, which formed the basis of a manuscript he was preparing for publication. Unfortunately, the following year, with his writing nearly completed, he passed away at the age of sixty-four.

Having studied under Atwood at Harvard, Peattie held one of the relatively few geography Ph.D.s awarded in America before the Second World War, and he gained a high level of credibility throughout the duration of his career, producing more than thirty publications before his death in 1955. He travelled widely in the pursuit of geographic knowledge, and may have been the only American academic geographer of his era that dared to venture into Sub-Saharan Africa for fieldwork. He was certainly one of the first academics in the United States to comment on Apartheid in a scholarly journal with "The Race Problem in South Africa" in a 1950 issue of the *Journal of Geography*.

Certainly, Peattie's passing created a marked impact on the department, which is reflected in an obituary in the *Annals of the Association of American Geographers*: "The Department of Geography and The Ohio State University have lost a devoted and distinguished member of the faculty. The loss of the prestige and distinction he brought is equally felt" (Smith 1957, p. 97). Although Peattie desired to be known simply as an author and a teacher, it is likely he served as an informal leader, giving much to the department and institution, as well as the geographical community as a whole.[3]

With the death of Peattie and the retirement of Van Cleef, Smith was tasked with rebuilding a department that had seemingly only been able to persevere through tumultuous times because reputable and conscientious colleagues were willing to share in the responsibilities necessary to achieve success. Given the high caliber of faculty, OSU's department of geography should have been by all rights considered on par with Wisconsin in the early 1950s, which then was likely perceived to be ranked among the top four in the country; however, like his predecessor, Huntington, Smith grappled with an administration unwilling to provide adequate monetary support to fulfill the needs of the department. Had the funds been made available, a distinctive graduate program might have developed more rapidly.

Regardless, both Huntington and Smith appeared to be effective leaders, and certainly not laissez faire, though not enough information is available to determine how much either embraced the transactional style and how much they incorporated transformational attributes into their leadership. Pieces of their stories lend support to the notion that transformational qualities were at times embraced. With regard to Huntington, his participation with colleagues in the International Geographical Congress in Paris, each at their own expense, speaks volumes about his willingness to set an example to facilitate professional development during an era in which transportation was cumbersome, costs were high, and this type of activity simply was not the norm. The fact that the National Council of Geography Teachers and Association of American Geographers met in Columbus in 1929, certainly with OSU geography department leadership managing the venue, also is indicative of the level of commitment the department chair placed on fostering relationships with the geographical community nationwide. Moreover, in the absence of other evidence, the level of scholarly contributions the faculty made under his leadership also is indicative of his support. Lastly, the scholarship fund that continues to bear his name, established by Huntington's widow after his death, indicates a high level of commitment as well; this also illustrates transformational leadership efforts.

Likewise, assessing Smith's leadership is difficult as well; however, the fact that he was chosen to lead the department after the retirement of Huntington, instead of the longer tenured Van Cleef or Peattie, suggests that his leadership qualities were recognized as superior to his colleagues. It is not necessarily his obituary that tells much, but the obituary he wrote of Peattie that indicates his grace and devotion to his colleagues. The unpublished history of the department he authored, which was updated by Taaffe, shows no evidence of aggrandizement of himself or the department, a criticism often levied against many of these studies. His history is balanced and provides context within the framework of conditions that influenced both the department and the institution.

Throughout the 1950s, twenty doctorates were awarded, several under Smith's guidance. It should be noted that three of Smith's most exemplary students, Arthur Robinson, E. Willard Miller, and H. Bowman Hawkes, became chairs in their respective departments of geography at Wisconsin, Penn State, and Utah. This graduate instruction occurred with monetary shortages, as well as a shortage of faculty; but Smith also had some of the finest geographers available on his staff, though they may not necessarily have had a true sense of camaraderie. Robinson reflected on Smith in an interview years later, "At that time the department was at loggerheads with one another and he was sort of a peacemaker. He remained the head for twenty, twenty-five years or more, I don't know. He built it up and so forth, but he wasn't a domineering individual. He was sort of a diplomat and he kept it going. It was a pretty good department by then" (Dow 1984).

Yet transformational leadership efforts do not always bring transformational leadership results. Smith's role in leading the department through difficult times was indeed significant, but like Huntington before him, he never appeared to be able to convince the administration to provide adequate support. By the end of the 1950s, the department had experienced several years of decline, and as a new decade approached, it was threatened with elimination. With Smith's retirement in the early 1960s, the College Dean offered the chairmanship to Ned Taaffe, a highly regarded geographer at Northwestern University, as well as the backing necessary to rebuild the program. Five new faculty positions and funds to offer competitive salaries did much to bring the department back from the verge of collapse (Gauthier 2002). With Taaffe at the helm, a new era in the department's history would begin, and Ohio State Geography would unquestionably become ranked among the elite.

NOTES

1. A special thanks goes to the Ohio State University Department of Geography for providing the unpublished history of the department authored by Smith and amended by Taaffe.

2. Peattie played an important role here, for he was personally acquainted with Richard Hartshorne, who was tasked with spearheading the geography division of the OSS. Peattie recommended to his colleague from the University of Wisconsin that his graduate student, Arthur Robinson, be placed with the intelligence agency as a cartographer. It was done.

3. Peattie asked before his death, "When you bury me, my friend, do not have stately mourning. Have no stranger pronounce platitudes over my body. Rather let some close friend read a grand and stirring thing such as I, in life, should have liked. Select some poem of adventure. Tell some tale of man's search for happiness . . . I shall have had my full drink of life" (Smith 1957, p. 97).

EIGHT

The Transformational Leadership of E. Willard Miller and Penn State Geography

As the country faced its early years of economic depression, the dean of the School of Mineral Industries at Penn State concluded that geography was an essentiality:

> Since the study of geography, with its emphasis upon correlations and interrelationships between the physical and cultural aspects of the world, is the intellectual bridge or catalyst between technical earth studies and human institutions, it becomes imperative that continued advancements be made both in the field of geographic research and in the field of mass dissemination of geographic knowledge through our institutions of higher learning (Miller 1992, pp. 134–135).

Thus, with the full support of the academic administration, the first professionally trained geographer was hired. In 1931, when Raymond Murphy arrived in State College Pennsylvania as an assistant professor of economic geography, he had already had several years of experience in academe.[1] Geography had not been addressed outside of summer sessions, and it was hoped that the new geographer could teach eight different courses in the College of Mineral Sciences, expand on the course offerings, and conduct research in the field of mineral economics with special reference to Pennsylvania. This would have been a difficult task for nearly any academic geographer; but Murphy's research and teaching background served him in good stead. He had completed a baccalaureate in mining engineering, a master's degree in geology, and had experience teaching undergraduate courses in both geology and geography. Within four years, a geography major was added in the College of Education, and in 1937, a second geographer, John Randall, was hired to teach courses on Europe, Latin America, and climatology; however, he remained on the faculty only until 1940, when he moved on to join the department at Ohio State (Johnson 1971; Miller 1992; Murphy 1932).

A greater variety of coursework in geography was now offered, including cartography and field methods, and Murphy also conducted research in urban and industrial geography, completing a number of studies. Among them was a regional survey of the state, which was published as the *Geography of Pennsylvania* in 1937. It appears that fourteen years in central Pennsylvania were enough for Murphy. In 1945, Murphy left for the University of Hawaii, and a year later took a position at Clark.

E. Willard Miller, who had recently been awarded the doctorate at Ohio State, arrived in State College as Murphy's replacement. "He was twenty-nine years of age and went into his Dean and said that he . . . 'wanted to make geography at Penn State the best in the world'" (Dow 1985). Geography had just achieved some autonomy as a division in the Earth Sciences Department, and for the first time was authorized to offer B.S., M.S., and Ph.D. degrees.

The demand for geographers was burgeoning in the federal government and other agencies, as well for faculty in public schools, colleges, and universities, so Miller developed a program that met their needs while also recognizing that service courses in the discipline were essential to provide students with an education that facilitated their ability in becoming knowledgeable public citizens during the Cold War. He wrote: "We need to spend more time in giving the average citizen sufficient information to discuss present-day world problems intelligently. So let us be assured of this, that the Russian question, the natural resources question, the South African race question, will never be solved until *all* people know a lot more about such basic matters" (Miller 1992, p. 138).

Miller had worked with Web McBryde and George Deasy in the OSS during the Second World War, and he was successful in adding Deasy, who was in the process of completing his dissertation at Clark, to the faculty in State College in 1947. He was an economic geographer interested in agriculture and mining industries. Phyllis Griess, who held a graduate degree from Wellesley, had taught during the War while also working as the dean's secretary, and was awarded the first geography Ph.D. in 1948,[2] remaining on the faculty until the mid-1960s. She also studied economic geography, with interests in Latin America.[3] Allan Rodgers (Ph.D. Wisconsin), an economic geographer with regional interests in Italy and the Soviet Union joined the department in 1952, and Frederick Wernstedt (Ph.D. UCLA), with interests in climatology and economic geography was added to the staff a year later. Thus, in six years Miller had convinced the administration to fund five faculty positions in geography that built the foundation for the program for many years. Success in serving both undergraduates and graduate students was rewarded with the appointment of two new faculty members. Peirce Lewis

(Ph.D. Michigan), a student of American landscapes, arrived in 1958, and Paul Simkins (Ph.D. Wisconsin), who became known as an outstanding university teacher, joined the program the following year, adding different perspectives in human geography. Shortly thereafter, the Penn State geography program was ranked eleventh place in a national ranking, a significant accomplishment for a department that was not in existence less than two decades previously, and Berkeley-trained Wilbur Zelinsky joined the staff further broadening the discipline's cultural focus (Brown 2004; Miller 1992; Paul Dean Simkins 2010).

Miller was proud of the work he had accomplished when he turned over the chairmanship in 1963 to Allan Rodgers (Dow 1985). He had every reason to be, for it was his vision and a collection of transformational leadership attributes that enhanced the development of one of North America's leading geography programs (DeVivo 2011). He worked tirelessly behind the scenes of American geography and accomplished much. His work in the successful merger of the elitist Association of American Geographers and the American Society for Professional Geographers, which resulted in a newly reconstituted egalitarian organization, was of paramount importance, as were his efforts to promote geography throughout his career. "There is no doubt that in the 1940s, Will was one of the young geographers who shook the existing disciplinary establishment to its core, to the short-and long-term benefit of all. He was wryly fond of noting that Preston James had once accused him of 'wrecking the discipline'" (Brown 2004, p. 665).

He never held a position as president of one of the leading national organizations; but he was recognized from coast-to-coast as the dean of Pennsylvania geography and as an icon in geographic education. Though not filled with philosophical discourse or diatribes, his scholarship was respectable and characterized by a solid simplicity that captured the essence of geographical field studies and research, whether it pertained to manufacturing or regional analysis.

As the program chair in the 1950s, "Will set the tone, style, and timetable for life and work in the department. . . . Graduate students saw him as they might see a Captain of a ship; someone who's responsibilities are great, a distinguished gentleman, and one at the pinnacle of the hierarchy" (Brown 2004, p. 663).[4] Though other faculty provided guidance to students in their own particular fields of interest, Miller was capable and willing to counsel all students.

> He was the overseer, giving time and energy to graduate and undergraduate students alike. He purposely provided only as much direction as he thought a student needed, thus creating a wonderful atmosphere for students who wanted to find their own niche in the discipline. His concern for students, and the discipline, could be gleaned by the brochure he produced. It spelled out with maps, pictures, and text, all that

> geographers would be capable of doing with a degree in hand. It was very inviting. . .
>
> Will ensured that students graduating from Penn State were as good as they could be. Both the workload and expectations were high. No wonder that, when the time came to advance the department toward the world of "big time" academics, Miller's high standards and strong work ethic provided an academic culture for the world-renowned program that emerged at Penn State (Brown 2004, pp. 663–664).
>
> . . . .in retirement, Will was twice as productive as any single faculty member who worked full time! You would see a steady stream of worried students going into Will's office, and an equally steady stream of students emerging with a sense of what they needed to do, when, and how. Will was generous beyond measure with his time.
>
> By his example, drive, and passion Will created a state-of-the art geography department. He would settle for only the best (Brown 2004, p. 667).[5]

Will Miller possessed the unique talent to inspire his junior colleagues, as well as his students, to reach for the stars and achieve *his* dream, which became *their* dream. Fueled with integrity, wisdom, discernment, and spirit, Miller established the foundation for Penn State to host one of the leading departments of geography in the world.[6]

NOTES

1. Awarded a baccalaureate in geology in 1923, he worked in industry for a brief while before returning to Madison to pursue graduate studies. Upon completion of the master's degree, he took a position at the University of Kentucky, where he was charged with teaching both geology and geography courses. Captivated by geography, he completed graduate coursework at the University of Chicago in the summer of 1928, and then accepted an appointment as a teaching fellow at the University of Wisconsin, completing the Ph.D. in 1930 under Vernor Finch. Murphy accepted the appointment at Penn State after teaching for one year at the normal school in Concord, West Virginia (Johnson 1971).

2. *University Microfilms* shows that the dissertation was completed in 1948; Miller's 1992 history of the program indicates she was awarded the degree in 1949.

3. The matter of Phyllis R. Griess deserves special attention, for women in academic geography outside of the normal schools were relatively few in number before the 1960s, and this was especially the case in the most influential programs. Certainly, Ellen Churchill Semple was an exception at both Clark and Chicago, as were Ruth Baugh and Myrta McClellan at UCLA. These three seemed to be the leading women in geography before the Second World War, for not only could they hold their own with most men in the profession, they could surpass many. This does not appear to have been the case with Frances Earle at the University of Washington, whose record was lackluster.

Regardless, perhaps what appeared most unusual about Griess is that although she taught some geography courses during the war, she engaged in secretarial work for the dean. Moreover, she was awarded the doctorate at the age of forty-eight, and then took on the third full-time faculty position in geography, after Deasy was hired. She must have made a deep impression on Miller, who not only likely served as a mentor, but also the department chair that hired her. This says a lot about Miller's leadership as well, for in going against the grain of "Old Guard" luminary geographers like Ivy League-bred Hartshorne and James in an effort to develop a more egalitarian Association of American Geographers was one thing (and he was harshly criticized for it); but to counter their views about women and seek to empower them in professional geography was quite another, for with few (often begrudging) exceptions, the discipline was considered the domain of men, not women. Here is just one example of Miller's sense of integrity and courage.

Incidentally, the 379-page dissertation written by Griess, *The Tin Industry of Bolivia: A Geographical Interpretation*, evolved into an article titled: "The Bolivian Tin Industry," published in *Economic Geography*, a journal then under the editorship of Raymond Murphy. Not at all an encyclopedic commentary as might be expected, it actually shows deep insight and raises some provocative questions about economic development, politics, and society at the time if its writing. It should also be noted that Griess co-authored a number of publications with Deasy, and also served with him as editor of *The Professional Geographer*.

4. Although Will Miller was indeed respected, it was known that not many faculty he hired harbored much good affection for him.

5. As I contemplate Will Miller's leadership, I recall sitting with him and his wife, Ruby, at Pennsylvania Geographical Society conference banquets years ago. We talked a number of times about geography in Pennsylvania, and it was obvious by his manner that he was no power-wielder; yet all who came in contact with him, spoke highly of him. He appeared to be always kind to students that nervously presented their papers in sessions before audiences of sometimes overly critical professors. If he was able to offer input, it was done in a way that was not condescending, nor disheartening, but helpful with a keen insight that was unusual for an elderly gentleman with his wife by his side. I learned at these conferences that Will Miller's presence was a given, for he essentially founded the society, and advanced geography not only in the state, but across the nation. At the time, I was not really aware of his contributions to the discipline beyond the vast sums he donated for geographic education, and I was surprised by the level of respect to which he was accorded, as he and his wife strolled slowly through conference hotel parking lots, corridors, and ballrooms. Relatively few geographers have served the discipline since the end of the Second World War with such devotion.

6. In 1995, the National Research Council ranked the geography department in State College as first among all programs surveyed (DeVivo 2011).

NINE

George Cressey and Preston James at Syracuse

By the time of the Second World War, Syracuse University hosted one of the leading departments in the United States for the completion of a highly regarded master's degree in geography. This was in no small way due to George Cressey, who had arrived as chair of the Department of Geology and Geography in September 1931. The son of the first woman that graduated from the University of Chicago, Cressey was bound to pursue graduate studies there; he completed the Ph.D. in geology under Salisbury in 1923. As he was influenced by his father, a Baptist minister, Cressey decided to work with the American Baptist Foreign Mission Society and teach at Shanghai College. Spending most of the 1920s exploring 30,000 miles of China, he learned much about the way of life in this remote part of the world, largely unknown to Westerners. At the end of the decade, he returned to the U.S. to complete a Ph.D. in geography at Clark.[1] Thus, with two doctorates he took on the role of heading the department at Syracuse, and it is little wonder then that those students completing their M.A. degrees were required to write theses "that took on the dimensions and intellectual qualities of miniature doctoral dissertations" (James 1964, p. 256).

J. Rowland Illick commented on the rigor of the program:

> When I attended Syracuse as an undergraduate there were only six majors in the whole discipline out of a total of six thousand students and probably twelve graduate students in 1940, 1941. Syracuse did not, at that time, have a Ph.D. program; so they tried to make their master's program the best in the country. George Cressey used to say it was sort of a small Ph.D. He had accumulated a good faculty and our requirements included (no language requirement) an eight hour written exam and the writing of a thesis based on fieldwork. . . . I finished the thesis

> in 1941 and considered it a very substantial piece of work compared to the other schools, where a couple of papers and some courses were offered, but no thesis was involved. I thought that this was good training for what I was to do, which was a Ph.D. at Harvard (Dow 1985).

Thus, Cressey was known as quite a taskmaster; but he also sparked an interest among students in studying the world around them and assisted them in their life goals, maintaining a degree of devotion that is often admired, yet rarely emulated, especially in large research institutions. Theodore Herman (1965) remarked: "His energy, his forthrightness, his concern for others, and his constant interest in learning touched all who knew him. He was indeed a great teacher" (p. 362). Illick also related how Cressey became influential throughout his life, recounting:

> I had a teacher by the name of Mrs. Cressey in the Shanghai American school who taught me history. When she learned that I was going to Syracuse she said I must look up George Cressey, who was her cousin; he taught geography there and she thought that I might like that field. When I got to Syracuse it turned out that he was my freshman advisor because I had come from China; and he was interested in China and I was assigned to him. That was the beginning of a very long friendship that helped me enormously, because he was a mentor for me and helped me to get to places like Beirut and Middle Eastern institutions where I taught for a while. He was always very supportive of the things that I did. . . . George Cressey was one of the major turning points in my experience. In terms of how he taught, his dynamism, his great enthusiasm for the world at large and, particularly, for his interest in Asia, because that was my first interest.
>
> Syracuse had an annual competition to see who was the most popular teacher on campus and usually it was Eric Faigle, another geographer. But in my judgement [sic], it was George Cressey. He was challenging, interesting, enthusiastic and excellent teacher. He had done his fieldwork, knew his stuff and knew how to present it. He lived it in class. All of these things really convinced me that I wanted to be more like the way George Cressey presented geography (Dow 1985).

While other geographers saw wartime service in the OSS, Cressey's expertise in the regional geography of East Asia sparked the Department of State to send him to China in the middle of the war. He also worked as a consultant for the Bureau of Economic Warfare and Army Intelligence. Known to discuss the inadequacies of American diplomatic efforts in his "characteristically outspoken" manner, he was later targeted by Senator McCarthy's anti-communist attacks. Fortunately, these were counterbalanced by the enemies lists produced by Chinese Communists (James 1964, p. 256). Theodore Herman (1965) would later comment, "Cressey knew well the disfavor he courted, but he never hid the truth about the achievements of others in order to feed public prejudice or display his own 'political correctness'" (p. 361). It was Cressey's integrity and willingness to speak the truth, despite virulent opposition and intimidating tactics launched against him, which characterized a strong foundation in his leadership.

In 1945, the binomial department separated, and as Cressey became chair of an autonomous geography program, he searched for a new faculty member to join him. Preston James recalled:

> I was at Michigan all the time from '23 until '41 when I went into the Army. After the war (during the late stages of the war) George Cressey came to Washington and invited me to come to Syracuse. I took leave and I went out to Ann Arbor to see what their attitude was and strangely enough they weren't treating me as a returning hero at this point. I went to Syracuse and Chancellor Tolley was tremendously excited about what he was going to do with a new geography department. He offered Cressey and myself all possible financial support to build a major graduate study and research center (Dow 1970).

> Cressey knew the Dean of Liberal Arts (his name was Eric Faigle) and, also, the Chancellor of the University. These fellows had a lunch for me and put on the heat. They told a very fascinating story of how we had the complete support of the administration. Cressey said: "This is a great university. You can have an unlimited future here." Naturally I bit on that one and I've certainly never regretted it (Dow 1985).

With a strong graduate program in place, Cressey turned over the chairmanship of the department to Preston James in 1951, while retaining his distinguished professorship as Syracuse University's first Maxwell School Professor. James had already established himself in geography at the University of Michigan, seen service in the OSS, and had just been elected as President of the Association of American Geographers. James had strong ties to the discipline's traditions, having been educated at Harvard and Clark by some of academic geography's early leaders, and he carried out a number of regional studies in the U.S. and Latin America, while also devoting much time and energy toward the historical and philosophical foundations of geography.[2] He had taught a course on the subject while at Michigan before the war, and he directed a project in the 1950s that resulted in the publication of *American Geography: Inventory and Prospect* (1954), which was largely an assessment of the discipline's history. In a letter written to him by George Cressey, appealing to him to take the Syracuse faculty appointment, Cressey remarked, ". . . our prime concern in inviting you to Syracuse is in the development of a strong Latin American program, and in your contribution to geographic philosophy in general" (Martin 1988, p. 168).

There were few practicing geographers that had studied geographic thought in such detail as had James, and he wrote a book on the subject, which has since gone through four editions (Martin 2005). Wes Dow recalled in a conversation with James,

> I recall, when I was a student of yours, I had been at Rutgers before and had taken History of Geographic Thought. I went into your office for a session (I was hoping to get a substitute course) and I told you I had had the History of Geographic Thought. You looked at me and said: "Well, you haven't had James." I took it and didn't regret it for a moment (Dow 1970).

James not only served as chair of the department during the 1950s, he continued to contribute heavily to geographical scholarship, amassing an extensive publication record, and at the end of the decade, Donald Meinig, who would gain eminence in historical geography, later joined the faculty. It was the foundation laid by Cressey's leadership as chair, which was followed by the effective chairmanship of James, which brought initial recognition to the Syracuse program. By 1962, the department was ranked third in the country among its peers (Martin 1988).

In the final analysis, it was George Cressey, who displayed strong transformational leadership attributes throughout his chairmanship, and this set the tone for Syracuse to take the lead at the beginning of the 1960s. The evidence indicates that he was characterized by a high level of charisma, drive, and vision, which captivated others and compelled them to embrace and commit themselves to becoming stakeholders in creating one of the finest departments of geography in the country. James (1964, p. 257) commented:

> He was a brilliant lecturer, but in many ways he was at his best in small seminars, where his incisive questions and comments brought forth the best efforts of his students. In no other activity did he so clearly reflect the influence of Salisbury. He was always impatient with mediocrity and insisted on high standards of academic accomplishment. All over the world Cressey's former students occupy positions of leadership in the profession and pay homage to the memory of the man who set them on their careers.

Cressey's reputation as an honorable man, and the level of rigor he imposed upon students in their pursuit of academic excellence, certainly worked in geography's favor. Being able to forge strategic alliances benefited the department's development as well. The strong level of support from the administrative leadership that facilitated Cressey's effort was unusual, but also characteristic of other programs that were able to rise to a level of prominence. It certainly must have helped that another geographer on the faculty in the 1930s, Eric Faigle, had gained recognition as a superb teacher on campus (and distinguished alumnus) and later became Dean of the College of Liberal Arts (Powers 1979).

That is not to say that Cressey sought to engage in power-wielding, for the evidence appears much to the contrary. He was committed to high standards of integrity and scholarship, as is demonstrated by his steadfast resistance to McCarthyistic attacks on his pronouncements concerning U.S. foreign policy, his rigorous requirements for students, and his own research largely derived from prolonged field studies in America and abroad. It was often these experiences, recounted in the classroom, which captivated students, convincing them to learn about the world beyond merely an encyclopedic inventory of facts. And it was his credibility as a scholar and a leader that sparked the institution to commit to building a program under his direction.

In contrast, though Preston James carried with him some transformational leadership attributes, he appears to have employed transactional leadership strategies with some frequency. His scholarship, characterized by collections of facts and ideas, rather than innovative concepts or initiatives, was perhaps even longer-lasting than Cressey's; moreover, he is recalled by many students of the discipline as a luminary.[3] Yet James was indeed a power-wielder on the national scale.[4]

James came from an affluent background. His father was a banker and stockbroker, and his mother was well-educated; it was a given that he would attend Harvard (Martin 1988). He was awarded the Ph.D. at the age of twenty-four, and when elected to membership in the Association of American Geographers at the age of twenty-seven, he joined a rather exclusive organization.[5] Within this context, it should be noted that the association denied membership to U.S. President Woodrow Wilson, who had been awarded the Ph.D. in history and political science at Johns Hopkins University, because he lacked "qualifications" (James and Martin 1978, p. 192). Thus, it should come as no surprise that James fought ardently against the merger of the Association of American Geographers and the American Society for Professional Geographers, and he would accuse Will Miller of "wrecking the discipline" because of his opposing viewpoints and efforts, which were directed toward opening up membership and conference participation to all (Brown 2004, p. 665).

Having said all this, it must be emphasized that James's scholarship and status in the discipline did much to keep the department at Syracuse a vibrant center of geographical excellence. He was loyal to his colleagues, students, and friends, and he worked diligently to advance the traditional study of the discipline. Hartshorne noted: "As one who has experienced with him a half-century in the profession, I can testify that during that period his role has been outstanding, his total impact second to none" (Martin 1988, p. 173).

NOTES

1. Although directed by Salisbury, Cressey's dissertation at Chicago, *The Indiana Sand Dunes and Shore Lines of the Lake Michigan Basin*, was also influenced by the eminent ecologist, H.C. Cowles, who closely examined this region via biogeography. Cressey's Clark dissertation on *The Ordos Desert of Inner Mongolia*, while addressing the aeolian processes that also were discussed in the Chicago manuscript, certainly

remarked on the tremendous loess deposits of the Asian region; this is evidenced in his subsequent publications on the subject. Given Cressey's record of academic accomplishments and solid fieldwork, it is little wonder that he showed himself to be a serious taskmaster.

2. James authored numerous works, and among them were textbooks in introductory geography and Latin America, which showed his talent for synthesizing information drawn from numerous sources into an organized framework.

3. Most likely have not heard of Cressey.

4. I recall being introduced to both Richard Hartshorne and Preston James at a wine and cheese reception sponsored by John Wiley and Sons in Washington, DC, during the 1984 Association of American Geographers Annual Meeting. I fondly remember shaking hands with them both, recognizing that they were important figures in the history of geography, but not realizing just how significant their roles were in the discipline's development. Although his vision by this time in his life was impeded by substantial blindness, at eighty-five years of age, James still commanded a notable presence; but to some he was filled with unwarranted conceit and he appeared to speak condescendingly toward others, especially those with whom he disagreed. For example, earlier that day, a session was held devoted to geographers during the Second World War. Gilbert White, a conscientious objector during the war, commented on his work associated with the American Friends Service Committee. Working with refugees in France, White had been taken prisoner by the Germans, remaining in captivity for a year until released through a prisoner exchange. As It became James's turn to speak, he declared (with a chuckle), "I disagree with Gil. I think it was a good war." He then went on to talk of the accomplishments of the intelligence services in Washington, DC.

5. Feeling indebted to his mother, who appears to have facilitated a sense of specialness within him, while also being raised with perquisites accorded to only the most privileged, it is not unreasonable to speculate that James was characterized by notions of entitlement typically associated with narcissistic personalities. His rapid rise in acquiring academic credentials, as well as prestige in the discipline, in addition to the ease in which he gained an Army officer's commission, lends evidence as well. Post (2004) has commented profoundly, and in great detail, about this phenomenon. Given that, the impassioned resistance James showed against the merger of the Association of American Geographers and the American Society for Professional Geographers is understood. By the same token, it must again be emphasized that his loyalty to his colleagues at Syracuse was indeed significant. Perhaps at AAG meetings there was an air of superiority around him; but on campus, he was personable and helped his colleagues as much as possible.

TEN

The Transactional Leadership of Wallace Atwood and the Emergence of Geography at Clark

When Wallace Atwood, a geography professor at Harvard, was offered the presidency of a university forty miles away, geography became established as a graduate program at Clark; here the discipline's founding in 1920 was far more complex and problematic than anywhere else.

Atwood had studied under Salisbury at Chicago, receiving the Ph.D. in geology in 1903, and serving for several years as a geologist for the U.S. Geological Survey and itinerantly teaching in public schools and a junior college before accepting an associate professorship at his alma mater. Realizing that he would not be able to assume a leadership role there until his mentor passed on, he accepted an appointment at Harvard as professor of physiography in 1913; William Morris Davis had just retired and another notable geomorphologist on the staff, D.W. Johnson, had accepted a professorship at Columbia, creating ample opportunity for an aspiring professor to advance.

Shortly after his arrival in Cambridge, Atwood developed an association with Charles Thurber, an editor for the publishing house, Ginn and Company in Boston. Thurber had been awarded the doctorate in education from Clark University and was the first alumnus of Clark to serve on the Board of Trustees. Through this relationship, Atwood came to author a series of geography textbooks for the elementary grades. Concomitantly, his efforts to develop Harvard's geography graduate program met with resistance from the administration, and he let his views be known that he was open to prospects elsewhere. Clark provided Atwood with that opportunity, and he took it (Cressey 1949; Koelsch 1980; Van Valkenburg 1949).

Clark University had developed a strong reputation for its graduate program in psychology; its founding president, G. Stanley Hall, a noted leader in child study and psychology, expressed to the Board of Trustees his intention to retire in 1919 (Koelsch 2002). The Board sought a replacement like Hall, who would assume the presidency and direct the psychology graduate program. When a suitable candidate could not be found, the idea of adding a new discipline to the university was raised by Trustee Thurber; that discipline happened to be geography, and thus began a sequence of meetings between Atwood and individual trustees concerning his notion of fiscal management and graduate studies in geography. Hall supported the selection of Atwood and regarded his elementary textbooks highly.

> Atwood set forth the terms on which he would accept a formal offer. These included the establishment of a 'graduate school of geography' which would prepare geographers for government, business, and educational institutions. In order to take leadership from existing departments and elsewhere, Atwood demanded a budget of 20,000 dollars the first year and 'at least' 25,000 dollars per year thereafter, exclusive of the president's salary and of library funding for books, maps, and illustrations. As part of a broader strategy . . . graduate programs in biology, chemistry, mathematics, and physics, in which Clark could not compete with larger institutions, [were to be] gradually discontinued. The remaining resources available for graduate work would be combined 'to promote what we think of as a great geographical institute,' comprised of a Ph.D.-granting department of geography and 'supporting departments in history, economics, and education.' Atwood specified that the geography department should include physical, economic, and commercial geography and climatology (Koelsch 1980, p. 573).

Atwood's terms were met, and in 1920, the Trustees of Clark University offered him the presidency. Two weeks later he submitted his resignation at Harvard, informing the president that he had accepted the Clark presidency with the intention of developing a Graduate School of Geography.

It appeared to the trustees, that during the post-World War I period of financial difficulty that characterized higher education, the plan to add geography and deemphasize other existing programs was sound. Murphy (1979) remarked, "They selected geography because they wanted a field that was relatively undeveloped in American universities and that was not too expensive. Atwood accepted the offer with the understanding that geography would have special status on the campus" (p. 39); however, shortly after his appointment came chaos.

The trustees had not looked with scrutiny at the financing of a new program in geography, and they were unaware as to whether or not the institution could bear the costs. Certainly, Thurber, as president of the Board of Trustees and Atwood's editor, clearly demonstrated a conflict of interest when he advocated on Atwood's behalf; but the deal had been made, though with some question as to its legitimacy. Secrecy shielded the trustees, the outgoing president, and Atwood; however, without faculty consultation, upon his appointment, the institution's confidence in the new president was questionable at best. In a succession of interviews, Atwood attempted to belay fears that standing graduate programs would be eliminated, though he continued to cite the need for graduate study in geography and the support of businessmen in its development; yet he did not deny rumors of faculty reductions in other departments (Koelsch 1980).

Within months of Atwood's appointment, the Clark faculty established the Graduate Board to review graduate programs with special concern for the adherence to long-maintained standards. When a new university catalog was published, and it showed a "Graduate School in Geography" had been established with Atwood as its director, professors across campus were surprised. Not long afterward, the suspicions and fears of many turned to reality as Ph.D. programs in the natural sciences and mathematics were eliminated. Resignations in biology and chemistry, retirements in mathematics, a natural death in biology, and a suicide in physics fueled their decline within the first few years of Atwood's presidency (Koelsch 1980).

Though Atwood received some support for his agenda from some members of the faculty, the authoritarian and clandestine manner in which his plan was carried out did little to foster trust within the academic community as a whole; alumni and the American Association of University Professors (AAUP) launched an investigation, with good reason, and Atwood lost support of nearly the entire Clark community except for the Board of Trustees that had endorsed his hiring. The situation came to a head when the popular liberal socialist, Scott Nearing, delivered a lecture on campus and criticized the noted economist Thorstein Veblen. Just as Atwood entered the lecture hall, which was filled wtih faculty, students, members of the local community, and the former president of the institution, the derogatory comments were made; Veblen happened to be Atwood's brother-in-law, and he took Nearing's criticism as a personal affront. Losing his temper, he demanded that the lecture end immediately and directed the audience to vacate the premises. "The Nearing incident was like a match setting off a series of firecrackers, and the atmosphere of the university remained fairly noisy over the next two years (Koelsch 1980, p. 579).

Atwood was so consumed with his desire to establish a solid graduate program in geography he ignored much of his duty as the president of the institution. The AAUP concluded an investigation determining that Atwood was lacking in "tact, administrative wisdom, coolness and good humor under opposition, fairness in controversy, and ability to see other men's points of view" (Koelsch 1980, p. 580). These characteristics were deemed imperative to directing the path of a university such as Clark, especially given the major reorganization that had taken place. The AAUP also criticized the trustees, and in response, Atwood asserted that no arrangement had been made for the diversion of funds from already existing programs or the elimination of these programs to subsidize Atwood's desired "school of geography." This statement was contrary to the facts, as it was completely untrue.

Regardless, Atwood received unanimous support from the Board of Trustees, and though the number of graduate students in the School of Geography was far fewer than those in other academic departments, its image of success, however inaccurate it might have been, was sparked by a selected group of graduate students and a stream of notable itinerant lecturers. Preston James followed Atwood from Harvard to the institution in Worcester to complete his doctorate, and he subsequently accepted an academic appointment at the University of Michigan. Ellen Churchill Semple, a charter member of the Association of American Geographers (and its president) lectured for several years. Well-known agricultural geographers offered periodic lectures as well: Homer Shantz, O.E. Baker, and Curtis Marbut. Clarence Jones, holding the Ph.D. from Chicago, Elmer Ekblaw (Ph.D. Clark), and Sam Van Valkenburg (Ph.D. Zurich) served on the faculty in tenure-track slots (Koelsch 1988).

Atwood left much of the task of the administration of the university to the dean, and he turned most of his attention to the geography program. Yet, despite his efforts to build a geographical institute of world-class caliber, the graduate program at Clark was little more than an illusion of grandeur. Merle Prunty (1979) reflected on his experience as a graduate student in the 1940s and commented,

> I fault Clark considerably for its limited emphasis on research training generally, particularly through problem-oriented seminars for doctoral candidates. The doctoral candidates rarely obtained more research experience through seminars than did the master's candidates. To my knowledge, my seminar experience was not greatly different from that of students who preceded me at Clark. One has to wonder how the course of American geography might have differed, since World War II, if the Clark of the 1920s and 1930s had been more research-oriented and less teaching-oriented in its doctoral programs. A lot of talented people went to Clark; the group of which I was a member was loaded with ability (pp. 44–45).

In 1945, Van Valkenburg, having been on the Clark faculty since the 1920s, expressed his dismay at the low salaries, little financial support, and weakness in the geography program. He wrote to Atwood, "What we need is a new Director, new energy, and a new program" (Koelsch 1988, p. 43). Atwood relinquished the chairmanship, and the following year Van Valkenburg took over as director; though a number of changes in faculty took place, the program remained largely unchanged during his sixteen years of administration. Raymond Murphy (Ph.D. Wisconsin) joined the faculty in 1946 after several years of experience at Penn State. He assumed the editorship of *Economic Geography*, which was housed in the Graduate School of Geography and took over as the school's director when Van Valkenburg retired in 1962 (Johnson 1971).

Interestingly, little changed in graduate instruction at Clark between its founding in the 1920s and the 1950s; it remained relatively provincial, and though by the end of the Second World War, more Ph.D.s in geography were awarded by Clark than any other program, not all Clark doctorates were equal (Rugg 1981; Koelsch 1988). Clarence Jones's students fared better than those of other members of the faculty, including Atwood. Quantity did not necessarily infer quality, and despite their own views, the faculty at Clark were not perceived as being part of a "Top 10" program within the first fifty years of the discipline's existence at the Worcester institution.

The reality is that for much of this time, Clark geography was marked by mediocrity, and at least during the time of Atwood's presidency, it was characterized by weak leadership. Atwood had developed a sound record of scholarship as well as geographic education in the form of textbook production, to be sure. And he attracted large numbers of students to his classes; one of the reasons for this is because he was known as an overly generous grader. Perhaps he thought that his mission was largely to evangelize and promote the discipline, not to judge. Whatever the reason, the notion that his geography courses were easy did little to establish credibility for the discipline—or himself (Koelsch 1980). Moreover, his willful deceit and authoritarian manner as president alienated many. Instead of developing allies in sister disciplines, he disaffected faculty throughout the university and sparked little, if any, desire among them to collaborate with geography.

Atwood obviously exhibited transactional behaviors as he negotiated his terms in becoming president of Clark University, and though his intentions may have been good, that being the establishment of a geographical institute, the way he went about it reeked of dishonesty and despicability. It is this type of legacy, which tarnishes a program, that persists in academic institutions for years, and one that makes success for the leader's successors extremely difficult. In retrospect, it appears that Atwood very well may have done more harm than good. Granted, the institute he envisioned was established—in his mind if no one else's; but at what cost? The alienation of faculty, alumni, and members of the Clark community blemished the achievement of Atwood's dream. He either did not realize or did not care that regardless of however noble the cause, unethical action in pursuit of a moral goal negates the goodness of the goal sought.[1]

It must also be mentioned that Atwood deluded himself into believing that the Clark geography program was superior to Berkeley, Wisconsin, and Chicago, asserting that it was "on the verge of establishing leadership in this country" (Koelsch 1980, p. 582).[2] That leadership would not come to Clark unless (or until) a transformational leader emerged on the scene.

As far as Atwood's successors go, it should be noted that although the Clark program continued to struggle for its credibility in the 1950s, Van Valkenburg carried with him a selection of transformational traits that facilitated relationships throughout the university community (Warman 1977). It appears that his tenure as director of the graduate school of geography laid down a foundation that enabled the program to resuscitate itself and move beyond the image it had generated during the Atwood years.

NOTES

1. The manner in which geography was established at Clark was broadcast throughout academe; it could be argued that as a consequence, the trustees and academic leaders at private universities elsewhere in the United States subsequently viewed the discipline (and its practitioners) with suspicion, and establishing geography at these colleges and universities would therefore not be considered. This was certainly the case at Johns Hopkins University when Isaiah Bowman was hired as the institution's president; he was forced to agree not to create a geography program. Yet he did so in an indirect manner, by using his influence with the White House and federal agencies to require instruction in geography for military officer training programs.

2. The noted (former) Director of the Center for the Analysis of Personality and Political Behavior for the CIA, Jerrold Post has concluded: "The more unrealistic the expectations, the more likely we will find a narcissistic individual driven by dreams of glory" (2004, p. 28). The narcissistic personality that characterized Atwood is evident, and only through the efforts of an informal business partner, who engineered his presidency, was he able to achieve the office he sought. Yet it was a facade many saw through, and the quality of a graduate education in geography at Clark was questioned. If more doctoral granting departments in the country existed at the time, perhaps the program at Clark might have met the same fate as the geography program at Harvard; however, the demand for academic geographers holding the doctorate obviously exceeded the supply, and as fortuitous circumstances fostered its emergence at Clark, fortuity enabled geography at Clark to survive long enough to achieve credibility as a solid center for graduate instruction.

ELEVEN

Duplicity and Deception at Johns Hopkins

When Isaiah Bowman assumed the presidency of financially strapped Johns Hopkins University in 1935, senior members of the faculty recalled what had happened at Clark University when a geographer was hired as the institution's president. They confronted Bowman and expressed their opposition to the establishment of a geography department because the prosperity of Clark's program came at the expense of other graduate programs in the institution, which suffered from the derailment of funds that were redirected toward geography (Koelsch 1980; Martin 1980). Bowman had done much for the discipline already ever since his days on the faculty at Yale. In 1915, he became the director of the American Geographical Society, an organization that depended upon large endowments to subsidize exploration, research, and publication. During World War I, he served as an advisor to President Woodrow Wilson, and he brought together a staff of extraordinary geographers to assist in the establishment of European boundaries at the Versailles Peace Conference. In 1931, he was elected President of the Association of American Geographers. "In their thorough search for a new president, the Board of Trustees of Johns Hopkins had sought above all an accomplished scholar, an individual who was a man of affairs, and one experienced in the art of raising funds" (Martin 1980, p. 181). One year after becoming the president of Hopkins, Bowman's face appeared on the cover of *Time* magazine to complement an article on financial difficulties at the institution (Education 1936).

Bowman served on a number of boards for government and private industry including the directorate of American Telephone and Telegraph, the commission to determine if the University of Puerto Rico should be established, the board of Visitors to the Naval Academy, and the Commission on the Structure of Maryland State government. He also consulted with a number of German geographers at the 1938 International Geographical Congress in Amsterdam, and discussed the inevitability of war. As an advisor to President Franklin Roosevelt, Bowman commented on the refugee problem, weather forecasting for military purposes, and geopolitics. He also traveled to South America to investigate the Peru-Ecuador border dispute (Martin 1980). Bowman's network of contacts in government and commerce in the United States and across the globe was unmatched. If anyone could bring financial solvency to Hopkins, it was Isaiah Bowman.

By the time the U.S. entered the Second World War, the university had expanded its military officer training program, which proved to be the key to positioning a geography program. In addition to other consulting roles for the federal government, Bowman advised the Secretary of State, the Secretary of War, and the director of the Army Specialized Training program. It was in this capacity that he was able to make a convincing argument for the inclusion of geography as an integral component of officer training programs. Subsequently, "If Hopkins wished to work with the army program, a geography offering had to be offered at once" (Martin 1980, p. 149). Not to do so would be considered unpatriotic, and arguably, by way of deception, Bowman brought geography to the Baltimore institution. Thus, a department was initiated in 1942, with Berkeley-trained cultural geographer George Carter appointed as chair. Carter was a student of Carl Sauer at the same time Bowman's son, Robert, was completing his doctorate under the dean of cultural geography at Berkeley.

An eclectic group of geographers joined Carter: urban specialist Jean Gottmann, historical geographer Andrew Clark, and Asian specialist Karl Pelzer. Both Pelzer and another notable Asian specialist with significant ties to the geography department, Owen Lattimore, had already been on the Hopkins faculty in the Page School of International Relations. "Bowman made the case for geography on the campus in terms of the national interest. While the faculty may not have been completely enthusiastic, the new department already offered more than a dozen courses, with other classes taught in the College for Teachers" (Martin 1980, 151).

Bowman conducted a strong campaign for funding geography, and when wealthy industrialist (and former General Motors vice president) John Lee Pratt joined the Board of Trustees, he gained an ally and was able to convince private corporations and foundations to lend support. The Isaiah Bowman School of Geography was established in 1948, and before 1950, climatologist C.W. Thornthwaite, renowned economist Ernest Francis Penrose, soil scientist Robert Pendleton, cartographer Lloyd Brown, and biogeographer D.H.K. Lee were added to the faculty. Though the future of geography at Hopkins seemed assured, a wrench was thrown into the works as the 1940s drew to a close. Bowman, who had retired in January 1949, could do nothing.

> When Owen Lattimore, who taught courses in the geography department, was accused by Senator Joseph McCarthy of being a Communist, and George Carter became one of Lattimore's adversaries, all that Bowman had hoped for was lost. A contemporary of Lattimore, E. Francis Penrose, recalled: 'Work was interrupted, Johns Hopkins was made the object of public attacks by witch-hunt organizations, and the effects were disastrous for the development of the department' (Martin 1980, pp. 156–157).

The quest for finding Lattimore guilty of establishing dangerous liaisons and threatening national security as "the top Russian espionage agent in the U.S." lasted for a decade and unfolded in national news organs such as *Time* magazine (Kahin 1989, p. 945). In the manner of a Shakespearean tragedy, protagonists capable of both good and evil created a marked impact not only on the discipline at Hopkins, but on academic geography in the U.S. as a whole. Here, Isaiah Bowman, George Carter, and Owen Lattimore played vital, albeit now largely forgotten roles.

When Bowman gave the institution's commemorative address in 1936, he remarked, "Freedom to think, to discover, and to report is more important than either Government or society itself" (Education 1936). Few exemplified this credo more than Owen Lattimore. Born in Washington, DC, the son of a highly regarded missionary and educator in China (who later retired as a Professor of History at Dartmouth), Lattimore was educated in China, Switzerland, and later England before returning to China at the age of nineteen. Lacking a formal academic degree of any kind, he engaged in work with an import-export firm, gaining considerable knowledge in the production, marketing, and flow of commodities in this part of the world, among the most foreign to Westerners, beset by social unrest and conflict. Unlike other Europeans or Americans that dared to venture into China's interior, Lattimore had no need of a translator; his command of the local vernacular and willingness to embrace local customs facilitated the development of relationships, which fueled a deep understanding of the lands and peoples of East and Central Asia. His first book, *The Desert Road to Turkestan*, published in 1928, provided an account of a journey that began two years previous; now in his late twenties, an accomplished author and explorer, Lattimore was invited by the Royal Geographical Society to deliver lectures, submit journal articles, and take membership. Because his work had attracted the attention of Isaiah Bowman, the then-Director of the American Geographical Society arranged for Lattimore to be awarded a Social Science Research Council fellowship for graduate studies in anthropology at Harvard, without holding an undergraduate degree. Field studies were subsequently conducted in Asia, sponsored by the Guggenheim Foundation, and articles were published in a variety of outlets, including *National Geographic.* Lattimore was offered the editorship of *Pacific Affairs,* the journal of the Institute of Pacific Relations, and after several months in New York City devoted to gaining familiarity with the position, he returned to China and continued his work as editor in Peking (Dunbar 2000; Harvey 1983).

By the time Bowman took the presidency at Hopkins, Lattimore, still with no academic degree, had established a level of scholarship, evidenced by an extensive publication record that was only rarely matched in academic circles. Thus, it should come as no surprise that when he wrote the new Hopkins president, inquiring about recommendations for prospective employment in American academe, Bowman responded, "There is only one university for you to consider with respect to the future and that is Johns Hopkins. I have had you on my mind ever since I came here" (Dunbar 2000, p. 29).

In the fall semester of 1938, Lattimore joined the faculty of the university's Walter Hines Page School of International Relations, and a year later accepted the position of Director; he continued to serve as editor of *Pacific Affairs* until 1941 and contributed to geographical scholarship through research and publication. Moreover, the White House and Bowman recommended him as an aide to Chiang Kai-shek, and he served in that capacity in China for several months during the early days of the War before being appointed director of the Pacific Bureau of the U.S. Office of War Information in San Francisco. In 1944, he was asked to accompany Vice-President Wallace on a tour of Siberia, Mongolia, and China, and when he returned to Hopkins at war's end, he was in demand as a lecturer and a writer for both the media and scholarly journals. Arguably, nobody in the English-speaking world at mid-century carried the knowledge and insight about matters Asian as did Owen Lattimore, and although he did not belong to the new geography department, headed by Carter, his work was substantially geographic in nature. Moreover, the interdisciplinary rapport practiced at the university tended to blur departmental boundaries.

Perhaps it was because Carter served as chair of the geography program, and for no other reason, that Lattimore was compelled to invite him to his home for a picnic in 1945. Harvey (1983) speculated that the invitation probably had been extended to "facilitate Bowman's pet project of a school of geography in close liaison with the Page School" (p. 6). Regardless of intent, Lattimore's decision to invite Carter to a picnic would have far-reaching consequences for the better course of a decade. According to Lattimore and three other guests, "the guests spent the day enjoying the Lattimore yard, touring the house, and eating hamburgers in midafternoon" (Newman 1983, p. 15). A couple of the guests were looking over the galley proofs of a forthcoming publication one had authored, and though Lattimore glanced at them, he appeared to be more concerned about attending to the weeds in his backyard.

Yet quite a different account was related to the Wisconsin politician whose reputation was built on a crusade to uncover Communists said to have infiltrated the U.S. Department of State. George Carter and his wife reported "clandestinely to Senator McCarthy in 1950 that Lattimore [and his other guests], acting suspiciously, conferred for a long time over some documents. Lattimore is alleged to have replied, that they were declassifying certain documents in favor of some friends" (Newman 1983, p. 15). Additionally, it appears that Carter also reported to McCarthy's office that on a trip to the Arctic Research Laboratory in Point Barrow, Alaska in 1949, Lattimore engaged in suspicious behavior (Newman 1983). These unfounded accusations, along with others from the virulent anti-Communist Alfred Kohlberg, resulted in several years of anguish for Lattimore, as he was placed on leave from his position at Hopkins while he underwent inquiry, interrogation, and inquisition; yet he managed to employ wit and humor as he engaged in rhetorical battle with his opposition.

One antagonistic witness, the former Communist, Karl Wittfogel, asserted that Lattimore recognized him as a fellow Communist in 1935, for he *smiled* when a journalist made the accusation. Lattimore responded, "If I smiled at all it was certainly a non-communist smile. Now I would be willing to believe that Communists have an arsenal of secret signals, but I would never suppose that it included anything as good-natured as a smile. . . . If I am wrong, and if a smile is a secret Red signal, I confess that I used to smile a good deal. In the pre-McCarthy days I used to think life was lots of fun" (Harvey 1983, p. 8).

By 1955, a lack of evidence led to the government dropping charges against Lattimore. He had withstood more than five years of harassment in public hearings that became issues in the 1950 congressional campaigns as well as the 1952 presidential campaign. Fortunately, his attorney was a man of inscrutable character; although affluent clients on retainer threatened to take their business elsewhere if he continued to represent Lattimore, Abe Fortas willingly lost much of his lucrative business by maintaining his commitment to civil liberties (Freeman 1982).

This is only one illustration of the impact the Lattimore affair had on American society, and it did not generate a favorable view for Johns Hopkins University. Perhaps it was because of a perceived tarnished image that outgoing president Isaiah Bowman in 1948 recommended to his successor, Detlev Bronk "to get rid of Lattimore" (Harvey 1983, p. 6). A dozen years earlier, Bowman assumed the presidency of the institution expressing the notion that freedom in academic circles was considered imperative, and two years earlier his remarks to incoming freshmen underscored the importance of integrity with a recollection of T.S. Eliot: "If there are no honesty and virtue in the heart there will be none in the home and if there be none in the home there will be none in the City" (Wrigley 1951, p. 41). Yet, it appears that a duplicitous Bowman quickly departed from pronouncements of probity when the false accusations against Owen Lattimore surfaced.

A consummate politician and power-wielder, it seems that Bowman saw nothing wrong with his hypocrisy. The FBI interviewed him some months after the famous picnic at the Lattimore home, and perhaps this federal investigation sparked questions in his mind about whether Lattimore's presence at Hopkins was worth the cost. Moreover, Bowman's relationship with George Carter appeared to be relatively close, which was facilitated by mutual admiration and shared kinship in geography; in fact, when Bowman passed away, it was Carter (1950) who authored his obituary for the *Annals of the Association of American Geographers*. Unfortunately, a record of their conversations is not available; however, it is likely that the graduate school classmate of his son swayed the views of the university president *vis-a-vis* the matter of Owen Lattimore.

What drove Carter, as the geography department chairman, to engage in a subversive campaign to illicitly tarnish Lattimore's image is a matter of speculation. Perhaps he truly believed Lattimore was a Communist; or maybe he was simply jealous. Was it perchance envy that also led to the denial of tenure during the last year of Bowman's presidency for Jean Gottmann, an extraordinary urban geographer who would retire with distinction from Oxford years later?

Carter had completed a Ph.D. under Carl Sauer, who had developed a reputation for exacting grueling standards; Sauer was quite the taskmaster, as students awarded doctorates from Berkeley arguably would have been able to complete Ph.D.s elsewhere much more quickly. At Hopkins, Lattimore, with no academic credentials whatsoever, was apparently viewed by some members of the faculty as being intellectually inferior, and this group appears to have included Carter. Regardless of Lattimore's extended field studies, publications, and highly regarded work for the government during wartime, *sans* Ph.D. he was not considered a credible academic by some of his colleagues who had followed traditional paths to the professoriate; for Lattimore to hold the position of Director of the Page School, guiding graduate students at the prestigious Baltimore institution was perceived as an affront to their profession. That the White House had chosen him to serve as a wartime advisor to Chiang Kai-shek seems to have added fuel to the fire.

Fortunately, with allies in George Boas, the head of the philosophy department, and Sidney Painter, head of the history department, testimony upon testimony was delivered verifying Lattimore's academic veracity, and 40,000 dollars was raised to pay for his legal counsel. In the aftermath, because President Bronk had not terminated Lattimore, but placed him on a paid leave of absence, Lattimore returned to the faculty. He commented about the reception he and his wife received in an interview with the *New York Times*, "The whole faculty at Johns Hopkins was terrific and they gave us a wonderful ovation when we came back. It was almost unanimous support and made us feel very good" (Breit 1950); however, the Page School folded and Lattimore was assigned to serve as a lecturer in the History Department (Newman 1983).

Subsequently, Carter became *persona non grata*, once his role in the affair became known, and he was compelled to step down as the chair of the geography program in 1958, the same year the Association of American Geographers honored him with an award for *Meritorious Contributions*. The *school* of geography had devolved to a *department*, and M. Gordon Wolman, a noted fluvial geomorphologist with a Ph.D. from Harvard (whose father had been on the engineering faculty at Hopkins) took over as chair. Wolman invited Lattimore to lecture in the geography department, and he did so quite ably, as he had done in the years before his suspension during the McCarthy hearings; Carter retired from Hopkins some time later and took a professorship at Texas A&M University. In 1963, Lattimore left Hopkins for Paris and a position at the Sorbonne before being appointed the Director of Chinese and Mongolian Studies at the University of Leeds in Britain, a position he held for several years. Never having received an academic degree, he was awarded numerous academic honors, including an honorary doctorate from Glasgow University.

In the history of academic geography, the development of the discipline at Johns Hopkins was most unusual and bore with it far-reaching ramifications, for it revolved around three individuals, all who took on formal leadership roles: Isaiah Bowman, Owen Lattimore, and George Carter. Had anyone else but Bowman assumed the institution's presidency, geography likely would not have emerged at all. By the same token, it is conceivable that the manner in which geography arrived raised enough concern and not enough support to see its status as a *school* survive past the tenure of its namesake.

Certainly, it was through Bowman's influence that Lattimore was brought in to direct the Page School of International Relations, which essentially served as the de facto geography program at the university before the Second World War. Not enough is known about his formal leadership in this capacity to draw legitimate conclusions about his leadership behaviors in higher education; however, Lattimore's editorial position with the Institute of Pacific Relations, his dedication to generating solid scholarship and journalism devoted to foreign affairs in Asia, and his role as a diplomatic liaison suggest that he was characterized by more transformational attributes than transactional, for it was through these efforts that he was most able to affect the beliefs and behaviors of a wide audience of followers, beyond the graduate students and colleagues he interacted with in the institution. Other elements of his personality point in this direction as well; for example, quite unlike other Europeans or

Americans of his standing, his engagement in dialogue with all members of a substantially hierarchical society (without benefit of a translator), as was the case when he explored Asian cultures and landscapes, indicates a willingness to facilitate relationships and understand the needs of common people. This type of knowledge gained was not intended to provide monetary benefit or grandeur; instead it was driven by intellectual curiosity and a desire to contribute to humanity by imparting sound counsel to those in government, both domestic and foreign, tasked with making decisions of consequence.

It also was through Bowman's influence that geography was formalized into a legitimate field of study and George Carter was brought in to the university, taking on the role of chair shortly after his arrival. Not much has been documented about his leadership as the head of the geography program; he had conducted some credible scholarship and guided some graduate students, but his interaction with colleagues and subordinates is not well documented. The denial of tenure for Gottmann and, most especially, his role in the Lattimore affair, however, are significant embarrassments, as is the devolution of the program from a *school* to a *department*. It's fair to say in retrospect that George Carter experienced more significant failures than successes while serving in a leadership capacity at Hopkins, and though he was honored by the Association of American Geographers, he departed in retirement without honor.

As for Bowman, he appeared to be the epitome of the transactional leader while serving as an advocate for the geographic discipline. Only through his influence with the White House was the requirement for geographical instruction addressed in the Army Specialized Training Program, and subsequently a formal department was established at Hopkins. He appeared to work tirelessly behind the scenes, engaging in a type of leadership oriented around contingent rewards and the exchange of political favors. He was effective, to be sure; but his efficacy in leadership lasted only as long as his life, which ended prematurely. The Isaiah Bowman School of Geography, which he saw established, folded at the time of his death.

Of these three leaders, it appears that Bowman is looked at with a mixture of views. He did indeed do much for geography—at the institution and for the discipline as a whole; however, he has been considered both a hero and a villain though he accomplished much during his presidency at one of the leading universities in America (Martin 1988; Smith 1987), it seems as if he could have done more as a leader, and his ethics have been viewed with suspicion.[1] Unfortunately, the question that remains unanswered is whether his intentions were to advance geography or to advance Isaiah Bowman. Moreover, once must wonder if his contributions to the discipline have withstood the test of time.

Carter is clearly seen as a villain of malevolent character. It appears that envy, likely rooted in insecurity and narcissism, drove him to do harm toward Lattimore. The validity of his scholarship has also been questioned, and this did not bode well for geography's reputation on campus.[2]

And Lattimore is the hero in this saga. Persecuted, but persistent, he fought the good fight against odds seemingly insurmountable and did as much for higher education and academic freedom as he did for geographical research and exploration.

NOTES

1. Why Bowman turned against Lattimore is open to conjecture; perhaps he feared being labeled a supporter of Communists. He was indeed overly concerned about his image and carried with him deeply set prejudices. In a conversation with John Boland, Professor Emeritus of the Johns Hopkins University Department of Geography and Environmental Engineering, he related to me that Bowman had developed a reputation for being a virulent anti-Semite to a degree that was considered even extraordinarily harsh for his day.

2. Boland indicated to me that on the Hopkins campus, no positive comments were ever heard in reference to George Carter. Further, it was widely suspected that he had fabricated archaeological artifacts to support his unique theories concerning diffusion. Consequently, the term, "Cartifact" came into being, which referred to these alleged counterfeit relics. To some degree, whether the artifacts were actually forged is irrelevant, for simply to raise the question and view the scholarship as possibly illegitimately fabricated brought disrepute to Carter, the geography program at Hopkins, and the discipline altogether.

TWELVE

Laissez Faire Leadership at Harvard and Geography's Demise

In the history of geography in North America, Harvard holds a special place, for it was here where the luminary, William Morris Davis, served on the faculty when he developed a theory associated with landform development that was embraced by geomorphology for many years. Davis also engineered the formalization of geography as an academic discipline with his influence cast in the meeting of the Committee of Ten in 1892, which affected the course of geography instruction in the public schools. He also played a definitive role in the founding of the Association of American Geographers in 1904, and the launching of its flagship journal, known in the vernacular simply as *The Annals*. Thus, he exhibited transformational leadership behaviors in directing geographic education, molding the organization of the discipline on a professional level, and shaping the paradigm that was to characterize geography during the period from 1892 to 1925.

Harvard also holds a special place in the history of geography because regardless of its ebb and flow at (or near) the top of national rankings, the university's status is unmatched. Thus, because it was here where geography suffered a humiliating loss in the termination of its program, the memory of Harvard as a Waterloo moment in the discipline's history remains, despite the role William Morris Davis played during academic geography's early origins.

By the end of the first decade of the twentieth century, however, his marked absence in the Harvard yard was notable. Here was an eminent geologist and geographer at America's crown jewel in higher education, and he apparently spent little time there in the years before his retirement in 1912. Koelsch commented, "Geography at Harvard, after a brilliant start in the 1890s was somewhat in the doldrums in that period, for although its recently retired Sturgis-Hooper Professor of Geology, William Morris Davis was a legend in American geography, he had been a rather elusive figure in Harvard classrooms for some years" (Koelsch 1980, p. 570).

Wallace Atwood, a professor in geology at the University of Chicago who exhibited fine credentials and a solid record was offered a position as Davis's replacement.[1] Appointed professor of physiography in 1914, he quickly became popular among students, partly because he was an easy grader. This "drew the attention of President Lowell, the chairman of Atwood's department, and at least two other colleagues, all of whom labored with him over the standards in his elementary courses" (Koelsch 1980, p. 571). Atwood also irritated the president when he lost control of his students on a field trip. So it should have come as no surprise that when he appealed to President Lowell for an expansion of the geography graduate program, not just once, but twice, he was denied both times. Disenchanted with his prospects at Harvard, Atwood left in 1920 to become President of Clark University and director of its graduate school of geography.

Robert DeC. Ward had been responsible for many geography courses, as well as climatology, which constituted his principal research, and after Atwood's departure he continued to take charge of these responsibilities. Ward delivered numerous invited lectures and amassed an extensive publication record while also committed to "teaching with a passion" (Brooks 1932, p. 37). He had been on the Harvard faculty from the turn of the century until his death in 1931 (Davis 1932). In addition to Ward, other geographical geologists and geographers that served on the faculty in the 1920s included Kirtley Mather, Kirk Bryan, Raoul Blanchard, and Derwent Whittlesey (Martin 1998). It was Whittlesey who made a great effort to develop an autonomous geography program, separate from geology; but instead, he unwittingly facilitated the discipline's demise.

Concomitantly, the wealthy wife of Alexander Hamilton Rice provided a significant endowment for the university in Cambridge in exchange for a professorship for her husband, a medical doctor and part-time explorer. An Institute of Geographical Exploration was established at Harvard with the approval of President Lowell; Rice assumed the directorship with no academic qualifications, and he also taught a course that required students to put forth little effort, which was largely oriented around recounts of various trips and expeditions (Dow 1985; Smith 1987). The highly regarded alumnus, Saul Cohen (1988), reflected, "A major failing on Whittlesey's part was to have accepted without protest the Institute's establishment in 1931. Strong objection should have been made then on the grounds that its financial patron and director, Alexander Hamilton Rice, clearly lacked proper academic credentials" (p. 149). Whether Rice attracted students to the discipline is doubtful; there was extremely limited interaction between the geography program, which was housed in an academic department, and Rice's institute. And even though the institute provided the esteemed cartographer, Erwin Raisz, with support and space for a cartographic laboratory and classrooms (Dow 1977; Dow 1985), the institute was probably more of a stain on the integrity of the academic discipline than a positive attribute. Quite likely, members of the Harvard faculty were embarrassed by the notion that a professorship could be purchased in the first place, and the idea that such easy courses were being offered to their students in the second. It could be reasonably argued that if this is what constituted the esoteric field of geography, it had no place in a higher education institution. Although Whittlesey's solid reputation for scholarship and professionalism spread across the discipline from coast to coast, it did not rebuke criticism from among his colleagues in Cambridge; there is no real evidence to indicate that at Harvard he was highly regarded at all.

When Whittlesey accepted appointment as an assistant professor on the Harvard faculty in 1928, he was already an accomplished scholar at the University of Chicago, and shortly after his arrival in Cambridge, he became editor of the flagship journal of the Association of American Geographers. He had contributed much to political geography and urban geography, to be sure, and "His role of friend and adviser to students paralleled that of teacher" (Ackerman 1957, pp. 443–445).The noted geographer, Peter Nash, commented that Whittlesey was "the wisest and most gentlemanly geographer I had encountered" (Dow 1977). Middlebury College Professor Rowland Illick recalled that while he was a graduate student at Harvard,

> The people I knew best were Whit and Harold Kemp. They lived in a small apartment just off the Harvard Yard. They had a very civilized way of living because they quit work every afternoon at 4:00 and had tea in their apartment. That was open house for anybody who wanted to come—the undergraduates and graduate students were always welcome in their home at that time. They would discuss current events, world affairs, almost any subject (Dow 1985).

And Saul Cohen, regardless of his criticisms of Whittlesey with regard to the matter of Rice, had kind words and admiration for his mentor at Harvard, as well as some of the geography faculty: "I benefitted immensely from the contact with Whittlesey, Ackerman, and Ullman. They were role models for me and, from their different perspectives and interests, an invaluable group of tutors. Sunday afternoons with other graduate students at Whit's . . . were events that revealed the dignity, keen intellect and essential humanity of this shy and retiring scholar" (Cohen 1988, p. 150). No doubt Whittlesey was successful as a teacher and a mentor.

He also was successful in seeing that his companion, Harold Kemp, was appointed to the Harvard faculty as a geography instructor, and this appeared to be his undoing. He had previously failed in his effort to gain approval for Kemp to join the faculty at Chicago, and presumably, Whittlesey had been smitten with his love for Kemp for more than a decade, writing to his family in 1916, "Nobody but Harold knows how much I like him and how hard a pull it will be if we have to separate" (Martin 1988, p. 152). Unfortunately for Harvard, as well as for geography, Kemp was, at best, a mediocre scholar and teacher. Moreover, he created significant embarrassment for the geographers at Harvard, as recalled by former students:

> "[H]arold Kemp, through a pattern of behavior . . . contributed very significantly to the problems encountered by Whittlesey and by geography at Harvard in the later 1940s. Kemp weakened Whittlesey, and ultimately and indirectly the cause of geography at Harvard by his deviant *public* (verbal abuse) behavior throughout an extended period in the 1930s and 1940s. In essence Kemp's personal statements in public, in countless one-on-one situations that went beyond just members of the university community, could only be described as execrable. His ongoing patent mental problems resulted in some of the most vitriolic outbursts directed personally at individuals we have heard in our lives. Not only did this behavior by an instructor in geography antagonize many, it also had become wide knowledge in the university even in the 1930s. For many in that community, Kemp figured prominently in their collective image of geography at the university. . . ." (Augelli and Patton 1988, p. 145).

Whittlesey appeared to be blinded by his devotion to Kemp and sought to have him promoted, claiming: "His teaching is superb. . . . If there are any brilliant geographers, Kemp is one of them" (Martin 1988, p. 153). No evidence supports these contentions; on the contrary, "Kemp's introductory human geography course was known as . . . 'a gentleman's C' course. It was anecdotal, gave little sense of what an academic discipline was all about, and did not convey the scholarly traditions of the field" (Cohen 1988, p. 148). It is unfortunate that a promising geographer was not hired instead of Whittlesey's companion, for had this been done, disaster may have been averted. Yet, the discipline's reputation had already been damaged by Atwood's attitude and low academic standards, Rice's absence of academic credibility; now Kemp's overall persona and pedagogy lent critics of geography corroborating evidence questioning its suitability as a course of instruction in a university.

In 1947, the administration at Harvard approved a proposal for a new associate professor in geography; however, the following year when the binomial geology and geography department recommended Ed Ackerman for the position, the administration informed the department that the new position was no longer available. Ackerman had been awarded three degrees from Harvard, achieving the grade of "A" in all his courses. He demonstrated solid geographical scholarship, and shortly after the award of the doctorate in 1939, though a member of the Harvard faculty, he was called to Washington, DC, for work in the OSS during the Second World War (Glick 1988; White 1974). Ackerman had drawn considerable support from a number of distinguished individuals who wrote strong letters of recommendation: Richard Hartshorne, Ackerman's wartime supervisor with the OSS; J.K. Wright, the Director of the American Geographical Society; Lt. Col. Hubert Schenck, Chief of the Natural Resources Section for the Supreme Commander for the Allied Powers (and who also served on the faculty at Stanford as a professor of geology) whom Ackerman worked with in post-World War II Japan; and Derwent Whittlesey, Ackerman's mentor. Even though a majority of the Harvard geologists supported Ackerman, the geology program chairman and the provost acted in collusion to eliminate the position that had previously been approved (Smith 1987). The president, James Conant, endorsed the provost's decision and declared "geography is not a university subject" (Glick 1988, p. 50). Faculty positions in geography were terminated; Ackerman left Harvard for the University of Chicago and Edward Ullman, who had done graduate work at both Harvard and Chicago before wartime service with the OSS, remained on the faculty temporarily while the fate of the discipline was relegated to a committee on university policy (Glick 1988; Harris 1977).

The committee was composed of faculty in economics, geology, government, history, anthropology, biology, and one geographer—Ullman. They met over the course of several months and deliberated over the nature of geography and its role at Harvard. Scholars from outside Harvard were called on to offer their views; Richard Hartshorne, L. Dudley Stamp, and Carl Sauer were among the witnesses that delivered oral or written testimony, as well as Isaiah Bowman, president of Johns Hopkins University, a Harvard alumnus, and an external member of Harvard's ad hoc committee that had recommended Ackerman be hired as an associate professor of geography.

Definitions of the discipline were as various as were the number of witnesses. Ullman expressed disagreement with some views of geography, most notably Sauer's, but also Bowman's. When the committee chair indicated that their views were to be included as significant components of their report, Ullman responded, "I am loath to include Sauer's letter in the report, as it stands. It represents very strongly his prejudices. . . . Neither Sauer nor Bowman is well informed on current trends in American geography" (Glick 1988, p. 57). In any event, the committee recommended that geography should be represented at Harvard as an independent department. Ullman claimed,

> For two years we fought and we wrangled about geography. They were very skeptical to begin with. At the end of two years they came around and produced a report which recommended unanimously the

> creation of a separate geography department at Harvard. A separate department had never existed; it had been in with geology. Well, this was a great moral victory and a moral vindication of geography by those who took the trouble to study it. Unfortunately, there were no dollars attached to this recommendation, so it's never been implemented. . . . that process had a great affect [sic] on me, because it made me want to prove the value of geography (Dow 1972).[2]

Discussion of Harvard's demise is not complete with addressing the role of Isaiah Bowman, for it has been addressed by a number of students that both studied and witnessed the termination of the geography program (Smith 1987; Augelli and Patton 1988; Burghardt 1988; Cohen 1988; Martin 1988; Smith 1988).[3] Bowman indeed played a part here, and he was certainly bothered about the Hamilton Rice part of the geographical equation at his alma mater.[4] He also appeared to be perturbed by Kemp's questionable academic qualifications.

Moreover, to say that Whittlesey's homosexuality was a significant disturbance to both Bowman and the Harvard administration is an understatement. It was known as an open secret in Cambridge, gossiped in the discipline, and considered an unacceptable behavior among academics, especially at Harvard at the time. Rumors of a "homosexual cell" involving a third geography faculty member ran rampant (Smith 2003). Geography at Harvard, which had toiled relentlessly to regain the ground it had lost since the departure of William Morris Davis, was now struggling to survive in a time and place marked by homophobia. By the same token, there was some validity to the allegations of nepotism and academic misconduct, as Whittlesey had brought Kemp to Harvard and sought to have him promoted. Interestingly, Bowman remained silent, and it is generally agreed that he believed that once a housecleaning at Harvard took place, he could be effective in the reestablishment of a credible program; but he passed away in 1950, and a leader in geography with his background, scholarship, and clout never reemerged.

Smith (1987; 1988) and Martin (1988) have debated Bowman's role in the demise. On the one hand, Smith paints a telling picture alluding to Bowman as the principal factor in the Harvard demise. On the other hand, Martin (1988) cogently asks, "But why blame Bowman?" (p. 155), and he offers an abundance of archival evidence that supports the notion that Bowman was supportive of geography at Harvard, just not Rice, Kemp, and Whittlesey.

Further, unlike many others that admired Whittlesey's scholarship, Bowman was harshly critical. He wrote Gladys Wrigley at the American Geographical Society in 1941 after reading Whittlesey's treatise on political geography, *Earth and State*: "A truly appalling book. . . . The style (or want of it) makes the book turgid beyond any possibility of developing interest. . . . I deplore its fundamental ignorance. It is filled with assertions that are just not true!" (Martin 1988, p. 155–156). Cohen (1988) delivered an unintentional rebuttal nearly fifty years later:

> The results of social science cross-disciplinary training speak more than Bowman's rhetoric. He was naïve when it came to Social Science. Whittlesey, educated in History and Geography, had a conceptual and methodological approach to the study of the national state that was powerful in its sweep, developmental in its approach, and meticulous in its design. When Bowman, the physical geographer, entered the world of human geography, the scholarly results were amateurish. Bowman's *New World* was an encyclopedia of facts thrust upon a base of physical geographic description. In contrast, Whittlesey's *Earth and State*, which looked at nation-building as an evolutionary process in harmony with spatial features and patterns, remains a landmark study in political geography (p. 150).

Quite likely, with few leaders in the discipline that embraced political geography, Bowman and Whittlesey among them, and each with different perspectives driving the field, envy and enmity appear to have been directed toward the gentleman at Harvard from the campus in Baltimore. And though when Conant's decision to terminate geography "was the presidential prerogative which Bowman respected" (Martin 1988, p. 155), his traditional viewpoint of the discipline led him to write to the Harvard president "I would not favor the establishment of a Department of Human Geography" (Martin 1988, p. 157), which was the direction Whittlesey was heading.

Having said all this, was Bowman responsible for the Harvard demise? Certainly not. Could Bowman have prevented it? Isaiah Bowman was the former director of the American Geographical Society, the president of Johns Hopkins University, an advisor to the White House, and one with official positions on a number of boards of private corporations and government agencies, ranging from AT&T to the Naval Academy. Bowman was the ultimate power wielder. It is difficult to imagine that if he desired to "save" geography at Harvard, it could not have been done. Yet Bowman was a transactional leader that engaged in cost-benefit analyses when it came time to make important decisions. The game theorist extraordinaire appears to have concluded that the matter of saving Harvard's geography program was not worth the cost.[5]

If not Bowman, then who was principally responsible for the demise? Even before President Conant pronounced the death knell for geography, Derwent Whittlesey had withdrawn from active participation in university affairs. "[H]e seemed to have withdrawn not only physically but also in spirit from the university's faculty and administration long before the crisis developed. Further compounding this attribute of Whittlesey's role was of course his poor administrative ability and his virtual collapse after the crisis came. . . ." (Augelli and Patton 1988, p. 146). Moreover, "Whittlesey was too far removed from the harsh realities of university political life to build and protect a department. His was the world of withdrawn scholarship. . . . Geography at Harvard in the post-World War Two years therefore lacked the leadership and the organizational strategy for assuming a lead role in the directions that American geography was about to take" (Cohen 1988, p.151). Essentially, though other factors perhaps contributed to the elimination of geography at Harvard, it was Whittlesey's poor judgment with regard to his support of an incompetent companion and his laissez faire style of leadership, which must be burdened with the lion's share of the blame. This is particularly disconcerting, for elsewhere in American academe after the Second World War, geography grew by leaps and bounds (DeVivo 2011).

When Ullman met with President Conant before he departed for the University of Washington in 1951, he commented that "he had a wonderful primary school geography teacher who taught him where the rivers and mountains were. 'That is why you abolished geography,' Ullman told him. 'We don't do rivers and mountains anymore.' Responding to the question of whether geography was a distinctive discipline, Ullman said, 'sufficiently distinctive and basic, and yet related to other disciplines in that it should be taught in very large universities'" (Glick 1988, p. 52).

The elimination of geography at Harvard has often been considered "'a terrible blow . . . to American geography' and one from which 'it has never completely recovered'" (Smith 1987, p. 155). Some students have speculated that geographical scholarship then was driven by methodologies, "which became devalued in broader mid-century academic discourse" and marked the signal "that elite research universities did not require geography departments" (Osofsky 2007, pp. 429–430). At Harvard, it was clearly a matter of questionable leadership in the geography program as well as in the administration throughout its history.

Yet the problem that peaked with Whittlesey was marked by antecedent conditions that extended from the time of William Morris Davis. He was indeed a leader in geographical scholarship, but his prolonged absences from the classrooms and campus during the latter stages of his distinguished professorship did little to attract students or create the impression of a need in geography beyond the strict confines of geomorphology. Subsequently, Wallace Atwood showed much promise when he arrived in Cambridge, but his overly generous grades had attracted the attention of the administration that seemingly saw this as corroborating evidence of the discipline's lack of credibility. And this was the administration that negotiated a deal for an endowment in exchange for a special geographical professorship granted to a medical doctor that traveled to unique parts of the world. Despite the accomplishments of Davis in geomorphology and those of Ward in climatology, these were the circumstances that Whittlesey was compelled to endure. Most definitely, his advocacy of Kemp on a professional level also contributed to the image of geography denuding in the Harvard Yard, and this appears to be the most significant factor responsible for the discipline's downfall; but the stage had been set before Whittlesey arrived on the scene. Albeit a gentleman and a scholar, he was not a leader, and a transformational leader as the chair is what Harvard geography needed most (DeVivo 2011). In 1956, a year before he was scheduled to retire, he passed away as the sole remaining academic geographer on the Harvard faculty (Ackerman 1957).

NOTES

1. Atwood contributed a number of articles on the physical geography of the Rocky Mountain region in addition to geographic education throughout his career.

2. Ullman had the opportunity to remain at Harvard after the geography program closed, especially since he held a half-time appointment with the urban planning program. He commented: "The city planners told me to come over on their staff full-time and promoted me so I could stay there, but I preferred to be in geography and went out to Washington" (Dow 1972).

3. On the one hand, it would appear that the role of an academic who happened to be the president of another American university would have been relatively inconsequential to affairs at Harvard; however, Isaiah Bowman was no ordinary academic, nor an ordinary university president. He was a Harvard alumnus that had gained a high level of distinction through cultivating a network of influential leaders in commerce, industry, government, and international diplomacy. Service to two wartime presidents, with whom he developed very close relationships, facilitated his ability to wield power and effect the course of events, and deliver much sought after counsel to captains of industry and foreign heads of state, such as Winston Churchill.

4. Bowman may very well have remained disturbed about the Hamilton Rice matter for personal reasons. When Bowman had been Director of the American Geographical Society, Rice's wife sought to purchase for her husband the Society directorship. The Society's Board refused to take the offer, but he logged the event in his memory.

5. It should be noted here that Conant also appeared to be envious of Bowman's status as President Roosevelt's confidante, as well as the taunts Bowman delivered. Brian Berry learned more than most about what happened "behind the scenes," when he took an endowed chair at Harvard in 1976. "When called to consult by Franklin Roosevelt, Bowman was always there. When Conant arrived, he invariably asked him about Hamilton Rice" (Berry 2006, p. 106).

THIRTEEN

G. Donald Hudson's Transformational Leadership at Northwestern

Before he arrived on the campus of Northwestern University in 1939, G. Donald Hudson had collected several years of experience in applied geography, teaching, and educational administration. After service in the Navy during the First World War, Hudson engaged in teaching at a rural school before enrolling at the University of Chicago. Subsequent to award of the baccalaureate and master's in education, he took a position as the director of the Middle School of the American University in Beirut. Departing Lebanon in 1929, he returned to the United States, becoming the dean in a small Illinois college before enrolling in the geography graduate program at Chicago, where he completed the doctorate in 1933. A year on the faculty at a Wisconsin teachers' college, and five years with the Tennessee Valley Authority (TVA), as an administrator charged with directing a number of geographers in land use planning activities, provided him with leadership and higher education teaching experience, as well as a background in the practical applications of geographical knowledge (Sherman 1991).

Although two others in the binomial department of eight faculty taught a few geography courses in addition to geology and the earth sciences, when Hudson was appointed professor of geology and geography, he was the only geographer on the staff and was charged with teaching geography courses exclusively. His teaching duties were reduced considerably when, in 1944, he served as Coordinator of the Navy V-12 Program, which was devoted to supplementing the number of naval and Marine Corps officers during the war. This experience provided rationale for his efforts to later elevate the status of geography at Northwestern, for during the War, the need for geographic knowledge was recognized as an essentiality, especially for military leaders.

> Angered by the fact that male students were not required to take courses which would help ready them for war conditions, Hudson made his opinions plain by singling out map interpretation and meteorology as prerequisites to national defense. . . . "There has been repeated mention in the press and from authorized sources on the radio that of all the fields in which the armed forces of this country need preliminary training map interpretation takes first rank. Meteorology is only second in importance. Is this not reason enough to make the taking of these courses a requirement? We owe it to the men, though they may not realize it. We owe it to the national effort; men in charge have expressed themselves" (Kenzer 1983, p. 33).

The seed for an independent geography department had already been planted in a memorandum to the president, and in a letter to the dean in 1943, Hudson stated the obvious need for such a program, and he outlined the plans for its implementation. The dean apparently responded favorably, because Hudson sent a subsequent letter offering greater insight. "There is not now in this country, nor was there in the year immediately preceding the war, a real center of geographic teaching and research. The department I have proposed would constitute such a center. Such a department is appropriate to my conception of Northwestern, and to Northwestern's position, and relationship to the Chicago area" (Kenzer 1983, p. 34). Inquiries were made to Ralph Brown, Richard Hartshorne, and Glenn Trewartha to determine their availability and willingness to take part in a new program at Northwestern. None of them appeared to be able or willing to do so at the time (Kenzer 1983).

Hudson continued to be persistent in his inquiries concerning the division of the binomial department and the subsequent autonomy of a geography program. Finally, his recommendation was acted upon by the administration, and in 1945 the separation of geography and geology at Northwestern occurred; the first graduate students enrolled in the new program, and Hudson would later be referred to as "Friend, Father, intellectual, and God!" (Kenzer 1983, p. 32).

He had assembled a solid core of geographers on the faculty. Edward Espenshade, the map curator at the University of Chicago, was serving in Europe with the Army when he received an inquiry from Hudson. He recalled:

> Don Hudson had set up a department at Northwestern and he invited about four or five people to come. I remember cogitating over this, when I was overseas, as to whether I should leave the University of Chicago and go to this "play school," as I had known Northwestern at that time. But it's difficult to tell why you make those decisions. Partially, my wife thought that she would like to be at Northwestern, because she had gone to school there, and so forth. I have never regretted the shift (Dow 1972).

Espenshade took on the task of editing the *Goode's World Atlas* through numerous revisions from the 1940s, and with newly arrived Clyde Kohn on the faculty, they both developed a greater emphasis on statistical analysis in geographical research. With Clarence Jones added to the staff as well, they engaged in the planning of the university's Transportation Center (Taaffe 1979).[1]

Jones had been on the faculty at Clark University for several years and served in the OSS during the war. He had developed substantial expertise in Latin America, having completed a number of field studies in Latin America for nearly two decades.[2] Wallace Atwood referred to Jones as "The best trained man in field geography that we have produced" (Hudson 1993, p. 168); it's little wonder that Jones took charge of the department's field camp for graduate students in Wisconsin, which required students to spend four weeks conducting fieldwork in both rural and urban areas, and then spend the fall term in a seminar, discussing their findings and putting the finishing touches on papers to be presented at conferences (Hudson 1992).

Thus the department at Northwestern integrated both the old and the new. It tasked students with applying geographical knowledge to resolving real world problems through research in the field, in the library, and in statistics and mathematics. By all means, Hudson was the catalyst that sparked the department's development in Evanston. By 1951, however, he had negotiated the chairmanship at Washington and moved on. He collected a sound cadre of geographers and provided the impetus for them to move the graduate program forward in the post-World War II period. Through his tenacity and will to get things done, he founded a reputable program. He commented, "My job as I see it is to do everything I can to create a situation in which members of this staff can most effectively perform their functions—teach and research" (Sherman 1991, p. 516). And that he did.

Jones took over as chair in the 1950s and continued to build on the foundation Hudson had laid down; however, despite his many accomplishments and the department's success, he was not the administrator that characterized his predecessor. Yet, it was under his leadership that the Transportation Center was founded, and his role was imperative as well, not only in his highly regarded research in Latin America, but also through his guidance of students in applications of field studies.

With the multidisciplinary Transportation Center in place, exemplary geographers were attracted to Northwestern. Among them was Ned Taaffe, who had completed a dissertation at Chicago, which was specifically devoted to the city's air passenger travel. Taaffe proved to be significant in his role as a mentor for students interested in the spatial analysis that required an integration of statistics with cartography. Richard Morrill (Ph.D. Washington) served on the faculty briefly, as did William Garrison, who had been awarded the doctorate from Northwestern in 1950 (Gould and Pitts 2002).

There had always been much collaboration since the founding of the department by Hudson, and this was evidenced throughout the tenure of Jones as well. The field course required of all graduate students drew from the perspectives of a number of faculty. Team-taught courses in spatial analysis incorporated the strengths and views of different professors as well. Grants from the National Science Foundation, which required the labor-intensive effort of a number of faculty, provided funding for summer institutes (Gauthier 2002; Taaffe 1979; Taaffe 1990).

Of significance was the collaboration with colleagues that followed similar interests through connections that were sparked in graduate school. With Ned Taaffe on the faculty and Brian Berry, who had taken an appointment at the University of Chicago, Garrison was joined at Northwestern with his former students from Washington: Michael Dacey and Duane Marble. [3]

> This Chicago-Northwestern center now became the Mecca for quantitative geographers in North America, the initial festival being the Symposium on Quantitative Geography held in 1960 and supported by the Office of Naval Research. This particular conference also introduced Torsten Hagerstrand to the North American movement, and the invitation was reciprocated with the participation of a number of North Americans in the Lund Symposium on Urban Geography held later that same summer. At the 1960 I.G.U. Conference which followed, the first Commission on Quantitative Methods was established with Garrison as chairman (King 1979, p. 127).

This type of cooperative effort was more unusual than it was the norm in academic geography. More often than not, individual faculty members chose to go their own way and pursue their individual research agendas in programs that demonstrated the eclectic nature of the discipline. At Northwestern, with the focus early on defined by Hudson and refined by Jones and Espenshade, the opportunity for dialogue about similar interests and collaboration in the achievement of departmental goals remained high. Certainly, Hudson had established a successful model; but it was carried out by a number of exemplary geographers that worked together to build upon that legacy. Jones, Espenshade, Taaffe, Morrill, and Garrison all served on the faculty within a decade of Hudson's departure for the University of Washington. All held leadership positions in academe throughout their careers, and four of them were elected to the presidency of the Association of American Geographers. For a relatively small department to make such a significant impact on the history of geography is a credit to the transformational leadership that continued from its founding, which was marked by the articulation of a vision and the implementation of the desire into a will to achieve.

NOTES

1. Raised in the rural countryside, Jones completed studies at the normal school in Macomb, Illinois, and served in the Army during World War I before completing the Ph.D. under Charles Colby at the University of Chicago. He supervised more than thirty doctoral students and was principally responsible for the development of Northwestern's Transportation Center and Taaffe's arrival as chair. Jones was known to be unassuming in character, despite the award of numerous honors and election as Association of American Geographers Vice-President (twice) and President. He considered the opportunity to award Sir Edmond Hillary with a medal from the Geographical Society of the City of Chicago to be one of his greatest honors (Hudson 1993).

2. Jones made his first visit to South America in 1925, which was funded by a Chicago industrialist, in appreciation for the work he was doing with boys' clubs in Chicago (Hudson 1993).

3. William Garrison had been awarded the Ph.D. at Northwestern and returned there in 1960, after serving on the faculty at the University of Washington.

FOURTEEN

G. Donald Hudson's Transformational Leadership at Washington

When Donald Meinig engaged in study at Georgetown University's School of Foreign Service, he picked up a copy of Ralph Brown's *Historical Geography of the United States*, shortly after deciding that a career as a Foreign Service officer was not his calling. He instead chose to pursue historical geography. Meinig had served in the Army during the war and conducted studies at Georgetown under the GI Bill; among other things, he noted that the witch-hunts of the McCarthy era tended to lessen morale in the State Department, which seemed to be a popular target for the Wisconsin senator, and his interest in pursuing a career in international diplomacy had waned considerably. Over the course of time, he would come to appreciate Brown's work as an excellent piece of historiography, but with a limited academic background in geography, he was on his own to pursue graduate studies and determine his own destiny (Meinig 1989; 1992). Thus, in 1948 he decided to return to his home-state of Washington and study at the university in Seattle, which he knew housed a large department of geography.

In retrospect, Meinig noted that his interests in the discipline were more akin to the kinds of things that Carl Sauer and his students were studying, but pragmatism compelled him to stay in Seattle. He commented:

> I had no advice whatever as to where to go to graduate school, but I knew there was a big geography department in Seattle, where I had taken two courses as a freshman and had actually talked to the chairman; and besides that, I think we were a little homesick for the West. It was not a very good department. Shortly after my time there, it was revolutionized under a new chairman and mostly new faculty, and became one of the most influential centers of a "new" geography in all

> the Euro-American world, but it was distinctly mediocre in 1948. Within a short while I realized that I should have gone to Berkeley, but practical reasons impelled me to persevere in Seattle. The not-very-taxing geography courses provided a solid foundation, and I read widely and roamed the campus in search of interesting lectures and courses (Meinig 1992).

Given the nature and caliber of Meinig's research and writing, it is no great surprise that he considered Berkeley a more appropriate fit.[1] Meinig's dissertation would be published as one of his several books throughout the years, and he eventually would be regarded as one of the leading geographers in the English-speaking world; but it's fair to say that he largely learned on his own, for his graduate training at Washington was indeed "mediocre."

Courses in geography at the University of Washington had been taught since the 1890s, and as was the case elsewhere, when a department emerged after the First World War, it shared binomial status with geology. The first professionally trained academic geographer arrived on the faculty in 1927; George T. Renner, had recently completed his doctoral dissertation on religion in tropical forests at Columbia, and increased the number of course offerings in geography. Albert L. Seeman also joined the faculty, completing his dissertation on economic matters associated with Seattle's urban geography in 1930 at Washington. That same year, Howard Martin joined the faculty, who had just completed his doctorate at George Washington University, studying economic activities associated with the Ohio River; in 1933, Frances Earle, also with a Ph.D. from George Washington was permanently added to the staff.[2] She had conducted research on geographic influences of British economic policy in Africa, and replaced Renner who left Seattle and served on the faculty at Columbia. In essence, after his departure, the department's members were focused more on basic economic geography than anything else, and the course offerings took on an increasingly regional focus. Phil Church joined the program in 1935, completing his dissertation on climatological matters at Clark two years later (Velikonja 1994).

The department's scholarly output was relatively insignificant before the Second World War. Seeman authored a physical geography textbook, which was one of only a few marketed in the United States at the time (DeVivo 1988); however, he was killed while serving as a naval officer. Like elsewhere in academe, the department contributed to the wartime effort by providing important instruction in military training programs, and faculty engaged in work with the government. Howard Martin had taken on the chairmanship when Renner had departed, and he found it difficult to secure qualified faculty, often relying on temporary appointments and graduate students. This was the case with John Sherman, who was teaching a number of courses during the war before completing his Ph.D. in 1947.[3] William Garrison (Ph.D. Northwestern) was hired the same year it was decided by the department that the geography program needed a change in direction to serve the needs of the burgeoning student body. Martin sought to resign his chairmanship, and Robert Hall, the chair at Michigan, and Fred Kniffen, the chair at LSU, were asked to consider taking over the department at Washington. They both declined; however, G. Donald Hudson, the chairman at Northwestern University, and already a notable figure in the discipline, accepted the appointment (Velikonja 1994). It would be the beginning of a new era.

Edward Ullman came from Harvard when Hudson arrived in 1951 and significant changes occurred. Some non-tenured faculty were quickly terminated, as they did not demonstrate the promise that Hudson had envisioned for the department, which had, as Meinig indicated, been characterized by mediocrity. A number of distinguished geographers filled visiting professorships, which supplemented the needs of the department, and a decision was made to concentrate the program's focus to four areas: cartography, economic geography, Soviet Union and East Asia, and the U.S. and Canada. Physical geography and cultural geography were not part of this new agenda, and the senior geographers, Martin and Earle, did not appear to conform to the plan (Velikonja 1994); they had been together on the faculty for twenty years by the time Hudson arrived, and they had been married to each other for most of that time. This created certain problems, as far as Hudson was concerned, and Earle was denied pay raises, contributing to tension within the program (Monk 2004). [4]

Essentially, Hudson had taken quick note of the university's strengths, planned and facilitated the development of a graduate program around them, and targeted capable geographers to fill much needed positions. Though the older professors on the staff seemingly were to some degree alienated, the department was now on a course headed for success. Garrison (1979) commented, "Sure there were some rough spots and conflicts and clashes among faculty stakeholders, but it amounted to a stimulating and useful activity, one that I have rarely seen so well guided" (p. 119).

Both Garrison and Ullman played pivotal roles in geography's "quantitative revolution," and they attracted a number of students that were interested in statistical applications to resolving geographical problems.[5] Though the department didn't entirely endorse this focus, it was supported by Hudson, both administratively and financially. He acted as a champion for his junior faculty and worked hard to accommodate their needs. "A high level of extramural contacts was supported by Hudson who was skilled at getting travel money for young faculty" (Garrison 1979, p. 120). Richard Morrill commented that Hudson "was not an apostle of the quantitative revolution but he understood what science was about and defended our overzealous endeavors along these lines" (Sherman 1991, p. 517).

Graduate students in addition to Morrill included: Brian Berry, Waldo Tobler, John Nystuen, William Bunge, Duane Marble, Art Getis, and Michael Dacey. They were all part of Garrison's famous seminar devoted to statistical methods in geographical research (Garrison 1979; Martin 2005; Velikonja 1994). Rarely in the history of geography in North America has there been a collection of such brilliant minds at one time in one place.[6]

Tobler commented, "Donald Hudson had the sense to put all the graduate students at desks together in a big room, and as a result of our interactions I learned just as much or more from fellow graduate students as from faculty" (Gould and Pitts 2002, p. 305). Yet, none of this would have been possible without Hudson in the role of department chair.

> He certainly did everything he could to assist Garrison and the gifted group of graduate students then in residence. He played the role of devil's advocate at times and stimulator and suggester at others. Since Donald served as both undergraduate and graduate advisor, he kept himself well-informed about graduate students' activities and interests as they related to those of the faculty. He was certainly a critical element in bringing to fruition the so-called "quantitative revolution" through his inputs to Garrison, the graduate students, and the departmental faculty at large" (Sherman 1991, p. 517).

Hudson realized the importance of having a solid group of faculty to achieve departmental goals. He arranged a visiting professorship for the Swedish geographer, Torsten Hagerstrand, who created a marked influence on the graduate students; but no one on the faculty appeared to make as significant an impact as did Garrison. In Gould's (1979) conversations with former graduate students that challenged an earlier paradigm, he asserted they claimed, "It was Bill Garrison who put his foot in the door and got us through" (p. 141).

Still, it was Hudson that set the foundation that made Garrison's work with his students possible. When asked by John Fraser Hart about what recommendations he would have for department chairmen, he responded, "Find out what you want to do, plan your route to that goal, then get the people that you need to implement your process; that happened at Washington" (Dow 1971). Though fortuitous circumstances somewhat played a role in their achievements, prosperity came to the Washington department partly through design. "As an administrator, Hudson was a marvel at getting things done, listening to and protecting young faculty, and promoting faculty interests" (Garrison 1979, p. 119).

His already established reputation facilitated his effectiveness in a number of ways beyond the University of Washington. Sherman (1991) reflected,

> He felt that it was his responsibility to search for appropriate academic positions for graduate students. He maintained voluminous correspondence with colleagues both in the U.S. and abroad—one purpose of

> which was to identify positions for which our graduate students might be logical candidates. His concern and work in placement was an important factor in the resulting widespread distribution, at major academic centers, of young men and women trained at Washington (p. 517).

This sort of far-sightedness enabled Hudson to achieve a vision of creating a world-class graduate program that would have a devoted following for an extended period beyond his tenure. By 1957, the program, which Meinig (rightfully) considered mediocre when he was awarded the doctorate several years before, placed seventh in a national ranking. In 1964, shortly after Hudson resigned from the chairmanship, the University of Washington's geography program ranked fourth in the country.

G. Donald Hudson exhibited the attributes of a transformational leader, and he employed these behaviors with effective results. The previous chair, Howard Martin, lacked the vision and desire to move the department beyond mediocrity, and it appears that he more-or-less embraced laissez-faire characteristics. When Hudson arrived, it must have been a rude awakening, for he did what had to get done to make things happen. Hudson boosted confidence and morale by providing financial and administrative support for faculty and graduate students; he inspired them to excel and achieve their dreams by articulating a clear and compelling vision for the future and supporting them to achieve their goals, which were consistent with the goals of the program overall. And he helped graduate students get jobs! Though Ullman, and especially Garrison, often receive much credit for the success of the Washington program, it was really Hudson, in many ways, who sparked the drive that fueled their success, and the success of their students.

NOTES

1. Interestingly, though Howard Martin, the department chair, directed his dissertation (Meinig 1953), Edward Ullman, who at Harvard had expressed disdain for Sauer's provincial study of geography at Berkeley (Glick 1988), signed his approval of Meinig's thesis as a committee member.

2. Earle was invited to come to Washington as Martin's temporary replacement while he was on a leave of absence in 1931. The following year, she was hired as an assistant professor of geography. Martin and Earle married in 1934, and she chose to keep her maiden name. Shortly after Hudson's arrival, he indicated to the Dean that her forthcoming retirement was expected. She remained on the faculty until 1967 (Monk 2004).

3. Sherman deserves credit for developing a program of cartographic excellence at Washington. He remained on the faculty until he retired. Velikonja (1994) noted: "John Sherman was part of the department from 1939 to 1986, for almost half a century. He was the founder of the modern cartography program at the university, the first full time cartographer at any American university, for over three decades one of the leading cartographers in the country." He was sensitive to the needs of the disabled and devoted teaching and research to creating maps for their use. For example, he tasked his students to ride in wheelchairs across campus, creating maps for the handicapped;

his wife elaborated on his application of cartography for the blind: "He taught himself Braille, put that on maps, and made Braille plates for elevators before the government made them mandatory. . . . He had some blind students, and he was so kind and gentle, he had a need to help them `see' as others did. His students just loved him" (Beers 1996).

4. Berry (2006) commented, "Hudson, I was told, was a stern taskmaster. His nemeses in the department were former chairman Howard H. Martin and his wife Frances Earle" (p. 41).

5. Although Ullman proved to be a vital element in the education of graduate students, "[he] was an imposing jowly man who smoked pungent cigars, liked to patronize, considered himself a gourmet, but wore his socks too long. . . . 'Big Ed' was very bright, but he was in love with himself, a congenital talker who was confident of his intellectual superiority and found graduate students a distraction" (Berry 2006, p. 49). On the other hand, it was Garrison that played the role of mentor. "It was apparent to all of us that Garrison was the man with the ideas. Despite his opaque classroom style (which he later admitted was deliberate), he was wonderful in small groups, especially one on one: funny, genuinely original and insightful, with those all-too-rare gifts of coming quickly to the nubbin of any problem and of inspiring excellence" (Berry 2006, pp. 51–52). Allen Scott, one of Garrison's graduate students at Northwestern corroborated Berry's views on his mentor in a conversation with me: "Garrison was a very inspiring figure. . . . He wasn't a very good teacher."

6. During a conversation with Waldo Tobler in 2010, I commented on this, and he modestly responded, "They brought out the best in us." And brought out the best they did. Each of them contributed much to geographical research and scholarship—and they were encouraged by each other, as well as their mentors. In a conversation with Brian Berry, he commented on the mentorship of professors and the "stimulus of an extremely bright cohort." I asked him, "Did you know how bright you were then?" He exclaimed, "Heavens no! We were full of our own uncertainties."

FIFTEEN

Iowa's Rise to Prominence

Before an independent Department of Geography was established on the Iowa City campus in 1946, Harold H. McCarty, who had served as a faculty member in the College of Commerce for many years, taught a number of courses in an unofficial geography "program" in the Department of Economics. Several of the courses had been part of the institution's effort to accommodate the needs of the Navy pre-flight training program located at the university. McCarty was assisted in this instruction by F.K. Schaefer, a faculty member who taught about the geography of Europe (Lindberg 2006).

McCarty had taken a Ph.D. in economics at Iowa and devoted much of his teaching to economic geography, with special reference to individual world regions, as well as political geography and the courses in the military program. He also conducted some credible research, which generated favorable reviews among noted geographers in the United States and Europe (King 1988; Lindberg 2006).

Thus, the department at Iowa was spawned by economics, unlike most other programs in the U.S., which were usually begotten from the earth sciences. As such, close ties were encouraged with the College of Commerce, and graduate students often took seminars in economics and sociology. McCarty (1979) recalled,

> Delay in its inception seems to have had a significant effect on the philosophy and perspective of the curriculum that was developed. Many of the young staff members who came to the department had taken graduate work in economics and sociology during a period in which debate was rampant in those disciplines over whether they should be approached primarily as science or as art. The former alternative seems to have triumphed and an attitude that social phenomena should be understood rather than merely described or appreciated

> came to prevail. That attitude seemed entirely applicable to geography and an analytical approach was favored (p. 122).

The faculty chosen by McCarty to become members of the department presented somewhat of a paradox. Robert Bowman, being the natural son of a famous geographer and the mentored "son" of another (arguably just as notable if not more so), was tasked with living up to the expectations of his bloodlines.[1] It could reasonably be presumed that Bowman would be a significant asset to the program and great things would be expected from this new assistant professor. On the other hand, Fred K. Schaefer, *sans* Ph.D., though an anti-Nazi refugee from Germany with some graduate study completed in Berlin, had shown no real promise in scholarship.

Looking backward, two things are clear with regard to these two assistant professors in the new Department of Geography. The one with the Bowman pedigree experienced a lackluster career. Although he was the son of Isaiah Bowman and had received his doctorate at Berkeley under Sauer, his courses devoted to cultural geography did not appear to attract a large following and his scholarship was negligible. Bowman left in 1950 for Nebraska, where he devoted the better part of his academic career to the university's extension division.[2]

Schaefer on the other hand, who showed very little promise, completed a written criticism of the then-current concepts in American geographic thought *a la* Richard Hartshorne, which was used by younger geographers, disinclined to accept traditional thinking, as a flag to rally around. He passed away before it was published in the *Annals of the Association of American Geographers* (Schaefer 1953); but that very well may have lent it greater credibility, for just as the death of Che Guevara has unified many in Latin America against capitalism, Schaefer united a new generation of geographers to band together against the traditional viewpoints of practicing academic geographers, many of whom were perceived to be inclined toward elitism. Bunge commented, "It was Schaefer whom the gang that Garrison gathered together at Washington pored over. Schaefer's methodology . . . launched Garrison and his raiders" (Bunge 1979, p. 131).

Hartshorne reflected on the episode by commenting on his experience visiting Iowa at the invitation of Schaefer (Dow 1986):

> I was invited by the department to come out for a couple of days for this purpose. We had one meeting. I think, the day before the seminar, an evening meeting at his apartment—his home. I met his wife and he had a number of colleagues there for drinks and food. My memory is that it was just general intellectual conversation—a good intellectual group. I don't remember any of the discussion. I think he took a minor role in the discussion. People asked me things about; wait a minute, I don't know what they asked me. I mean, I don't remember at all, except that it was pleasant. The next day was the seminar and that was very interesting, because he clearly had his students primed; I

shouldn't say he had them primed, but he had them taught to look critically and be ready with questions. I don't think I have ever had an audience that had so many good questions. Very early on I figured these people have done their homework. I've got to be on my toes to come through with good answers. He acted as moderator. I can't remember that he said anything himself, but just controlled the discussion; he did it very well. I thought at the end that was a good show. They were much freer to be critical than my own students at home. I encourage my students to be, but still you can't get them to be fully so. These people were really going after me, so to speak, but quite impersonally. All questions were perfectly proper and in order, and I got the drift. They objected to the comparison of geography and history. That's what they were attacking. In fact, one of the students had written a paper for the seminar, which she sent to me afterward. I read it and replied to a number of her points. I thought it was a pretty good paper and I turned it over to David Lowenthal, one of the students (a major in history) in my seminar. He wrote her a longer letter disagreeing with her view of history. Well, that was that weekend. A week or so after Schaefer wrote me a letter.

Dear Hartshorne:

Just met my seminar on methodology and, of course, the Saturday morning session was discussed. Without going into detail of the discussion, let me express my and the students heartfelt gratitude for that session. Naturally, there was disagreement here and there, but everyone felt the same about your fortitude and splendid response to the questioning.

Then it goes on and he concludes:
Miller, Bergmann, Moehlman and Horn told me what a treat the evening was. Again, thank you,

Yours sincerely,
F. K. Schaefer

To say that Schaefer's paper made an impact is an understatement. If it were not for his essay, there's no reason to believe that the role of Iowa's geography program in the history of geography would have been anything other than unremarkable; yet, it appears that McCarty not only was reluctant to support the publication, he discouraged any written endorsement of it at Iowa. Lindberg (2006) explained:

There is no question that Schaefer was a difficult colleague for McCarty and perhaps others. Schaefer's article made a number of statements critical of Hartshorne and the sources that he used in writing the highly regarded book, The Nature of Geography, 1939. . . . Hartshorne was a significant and respected scholar in American geography and McCarty was concerned that the young department not lose its claim to credibil-

> ity in the profession by supporting Schaefer against Hartshorne. McCarty was on the editorial board of the *Annals* at the time Hartshorne's lengthy rebuttal appeared . . . The editor at the time was Walter Kollmorgen from the University of Kansas and McCarty certainly knew the feelings of the editor. He probably also knew the feelings of other members of the *Annals* Editorial Board . . . (pp. 14–15).

Upon reflection, it is interesting that the program at Washington, which demonstrated substantial support for Schaefer, consistently has ranked in the "Top 10" departments in national surveys. Iowa never entered the "Top 10." It could be argued that Iowa's most noteworthy chapter, with regard to the history of the discipline, relates to the Schaefer-Hartshorne debate, and it is unfortunate that in Iowa City it was taboo. Perhaps even more significant is the notion that except for the iconic image of Schaefer, the University of Iowa would not be examined in this work devoted to examining the most influential departments of geography in the United States. It would merely be relegated to a minor role along with the majority of other doctoral programs in geography.

To judge McCarty solely on this issue, however, would be unfair, for he did much for geography at Iowa by investing himself in the department. Neither he, nor his program, was immune to the political realities of the time, and he seemingly employed transactional leadership behaviors to achieve the level of success his program gained. By the end of its first decade in existence, the department ranked thirteenth nationally, and it was considerably smaller, in terms of faculty numbers, than its peers (Lindberg 2006). Still, one cannot help but wonder if geography at Iowa might have achieved greater success if, like at Washington, Schaefer's views were discussed.

In any event, McCarty worked hard to elevate the program's image through his scholarly research and networking in the professional community (he was after all on the editorial board of the *Annals* and he published numerous articles); however, the matter of faculty hiring must be addressed. Of the seventeen faculty members that took tenure track positions in the nineteen-year period between the department's founding in 1946 to McCarty's resignation as chairman in 1965, only seven remained on the staff for a period of seven years or longer: McCarty 1946–1968, Hartman 1946–1956, Schaefer 1946–1953, Rumage 1955–1968, Salisbury 1955–1979, Kohn 1958–1979, and Lindberg 1960–2002. In a doctoral program, the low retention can be especially problematic for attracting students and providing continuity for their study. Though some attrition must be expected, the relatively high turnover rate of approximately 60 percent seemingly indicated that new faculty were either often poorly screened when initially hired or disgruntled by the academic climate and environment in the department. This represents a failure in leadership.

Regardless of the departmental climate for faculty, graduate students were indeed attracted to Iowa, just as graduate students are attracted to a number of other departments. Yet unlike other departments, dialogue and debate were productive in shaping intellectual minds prepared for comprehending both sides of arguments, not condescending diatribes so often the norm. Golledge reflected, "At Iowa, I learned the rough-and-tumble and give-and-take of academic interchange, but it was constructive debate and criticism, not the destructive type that is so common of much postmodern and 'critical thinking' in geography today. I took courses from Gustav Bergman (a member of the Vienna school and a logical positivist), who taught me a new and different way of looking at many problems" (Gould and Pitts 2002).

The continuity lacking among faculty in the geography program could be found with Bergman, whom nearly all graduate students relied upon in shaping their philosophy of geography. He had been a personal friend of Schaefer and seen that his landmark article was published in the *Annals* after his death. Additionally, as the 1950s wore on, within the department some stability was offered with appointments to the faculty of Neil Salisbury, Clyde Kohn, and James Lindberg. Each made significant contributions in scholarship, which was considered imperative for providing a solid geographical education (McNulty 1991).

In retrospect, Harold McCarty played a significant role in the founding of Iowa's Department of Geography. It was McCarty that shaped the course of the program by accentuating its economic focus. For a department to emerge as an autonomous unit at the end of the Second World War was an accomplishment in itself, let alone for one to materialize offering a credible doctoral degree. Fortunately, at Iowa, much was borrowed from ancillary fields and digested with a spatial focus, and for McCarty to make collaborative arrangements of that nature took special skill. Yet not much evidence exists to support any notion that McCarty exhibited transformational leadership attributes; on the contrary, his leadership style appears to have been more motivated by the politically charged wheeling-and-dealing of transactional leadership.[3] Given that, it makes one wonder if Iowa would have soared into the "Top 10" if transformational leadership behaviors had been employed.

NOTES

1. Robert was the son of Isaiah Bowman and mentored by Carl Sauer.

2. Bowman had been awarded the baccalaureate from Dartmouth in 1935. His doctoral dissertation was devoted to an examination of soil erosion in Puerto Rico, and during the 1940s, he spent considerable time conducting fieldwork in Australia, New Zealand, and New Guinea, seeing some of his research published in the *Geographical Review*. After three years at Iowa, he moved on to Nebraska, where he developed a reputation for alcohol consumption and irascible behavior. Edwin Hammond had been on the faculty at Nebraska briefly before going on to Wisconsin; he related to me in 1986 that Bowman's behavior was so egregious that Homer Aschmann, who had just accepted an appointment on the faculty there, was on the verge of leaving Lincoln

before the middle of the academic year. He did remain for the academic year and then departed, but Bowman's personality was contentious and it did not go unnoticed. He was forced to leave the geography department at Nebraska and moved to the university extension division.

3. Gerard Rushton reflected on his arrival as a graduate student at Iowa: "Iowa geography in 1961 was like nothing I had ever encountered. The small faculty (six) and graduate student group (about 30) were clearly on a crusade. They saw a new geography founded in philosophical positivism and implemented with quantitative methods" (Aitken and Valentine, p. 172). Given that, it would be unfair to create the impression that McCarty was merely a politician in academe exclusively devoted to contingent rewards, for he did establish an environment of provocative faculty that offered inspiration and alternative views on geography for graduate students like Rushton and Golledge, and this was no small feat.

SIXTEEN

Transformational Leadership at UCLA

The development of a geography program at UCLA was unlike any of the other leading departments in that the institution originated as merely a branch of the Normal School based in San Jose devoted to the training of schoolteachers in 1881; appointment of a full-time faculty member occurred in 1895 with the hiring of James Chamberlain. When departmentalization of the institution occurred in 1911, the Department of Geography emerged as an independent administrative unit. Shortly thereafter, Ruth Baugh and Myrta McClellan were added to the teaching staff; however, when the institution was reorganized as the southern branch of the University of California in 1919, Professor Chamberlain decided to leave because the restructuring mandated that faculty without the Ph.D. were to be demoted to the rank of instructor, regardless of experience or scholarship. Chamberlain had taught at the institution for twenty-five years before his departure in 1920, he had served as an associate editor for the *Journal of Geography*, and he would develop a reputation for his geography textbooks designed for schoolchildren. He eventually became the president of the New Mexico Normal School at Silver City (Dunbar 1981).

McClellan subsequently became chair; like Chamberlain, McClellan was deeply involved in geographic education in the public schools, having had several years of experience teaching in New Mexico, Colorado, and Illinois before completing the baccalaureate in geography, as well as a number of graduate courses at the University of Chicago. She had written to Salisbury in 1913: "I want to work with adults rather than children. I also wish to organize the work and execute it according to my own plan. . . . [A]s you know, I've been preparing myself for a number of years to be a teacher of teachers. . ." (Monk 2004, p. 7). McClellan maintained a high level of commitment to geographic education, evidenced by the teaching of many classes as well as fieldwork (Monk 2004). This was obviously noticed by her supervisors, for she was assigned to work in Sacramento for a year with the state board of education to integrate geography in elementary schools.

Because Derwent Whittlesey had taught in the summer of 1921 and created a favorable impression, McClellan made inquiries about whether or not he'd be interested in teaching in Los Angeles; but he showed no inclination to move to California. She contacted Isaiah Bowman, Director of the American Geographical Society in New York City. He recommended George McBride, a recent Yale Ph.D. that had been working as a research associate at the Society's office in New York (Dunbar 1981; Spencer 1972).

In the meantime, realizing that advancement beyond the rank of instructor was not possible in the university without the doctorate, Ruth Baugh conducted studies with Ellen Churchill Semple at Clark University. The luminary had taught as a visiting faculty member at the Los Angeles institution twice before it became part of the university system, and she attracted Ms. Baugh as a protégé; the instructor from Los Angeles journeyed to Clark to receive graduate training in geography for both the M.A. and the Ph.D. When Ms. Semple completed her last book, *The Geography of the Mediterranean Region*, it was Ms. Baugh who saw it to fruition after Semple suffered from a heart attack and was found incapacitated and unable to complete the work on her own. When Baugh completed her doctorate in 1929, she was rewarded with a tenure track appointment as an assistant professor.

Thus, UCLA was most unusual in having two women on the faculty in the same department of geography (Dunbar 1981). This progressive viewpoint on the part of the institution would not last forever; were it not for her long history with UCLA, Baugh likely would not have secured the permanent position. It would be decades before UCLA would hire another woman, and the meddlesome administration engaged in choosing new faculty without departmental input for several years (Dunbar 1981). Yet given that the times were marked by rare entry into the professoriate by women, Baugh, and especially McClellan, showed transformational leadership attributes in their roles among the initial faculty members at UCLA. Hallock Raup, one of Carl Sauer's early students and chair of the geography department at Kent State would comment on McClellan's impact in directing a number of students, including himself, to study geography, as he responded to an inquiry about driving forces that influenced him as an undergraduate:

> I think it was mainly an excellent teacher of the old environmentalist school by the name of Myrta Lyle McClellan. Very few people have heard of her, but she has been responsible for a good many of the

> professional geographers in the field at the present time. She was a product of Chicago. She didn't care at all about, research writing, or anything of the kind, but she was an excellent teacher, well organized and she is responsible for a lot of us going into the subject (Dow 1973).

Thus, McClellan appears to have been marked by transformational leadership behaviors, and when George McBride arrived in the fall of 1922, he would demonstrate evidence of transformational leadership as well. The administration expressed confidence in him early on, for the director of the Los Angeles campus wrote: "Dr. McBride is here and I have a very good impression. I have not had more than ten minutes opportunity to make his acquaintance but he looks and acts like a genuine person, modest, quiet, clear-cut, with a Kansas visage and a past experience of the world" (Dunbar 1981, p. 11).

McBride had served as a Protestant missionary in Chile and Bolivia for several years at the beginning of the century. It was in La Paz where he met Isaiah Bowman, who encouraged the missionary, a year shy of his fortieth birthday and "appalled by the squalor and hopelessness of the lives of the unpropertied," to consider a professional career in geography (Aschmann 1972, p. 685). McBride migrated back to the U.S. and began work first as a geography instructor at Wesleyan University, and later as an assistant librarian and research associate at the American Geographical Society, while pursuing doctoral studies in geography at Yale under Ellsworth Huntington. His dissertation was conducted, to a large degree, independently of Huntington, the noted environmental determinist, who sometimes forgot the appointments he had made with his student and on occasion could not be found (Aschmann 1972; Spencer 1972). [1]

It was under McBride's direction that the autonomous geography department at UCLA grew by leaps and bounds in the 1930s, and a master's program was incorporated into the curriculum. He resigned the chairmanship in 1942, when he accepted an appointment as an American advisor to the Ecuador-Peru Boundary Commission. Continuing in this capacity through the 1940s as well as holding the rank of professor at the university, McBride announced his formal retirement from UCLA in 1949, and during his career he had completed a respectable volume of scholarship for which he was awarded an honorary doctorate by UCLA and the David Livingstone Centenary Medal by the American Geographical Society. Reaching beyond the university, his research by itself was transformational in that it was translated into Spanish and considered an important influence in promoting progressive agrarian reform in Latin America (Aschmann 1972). Spencer commented on his personal relationship with McBride:

> As a teacher, a geographer, and a very human person, George McBride was gregarious and fun loving, but he never laid aside the sterner elements of his early education. When affairs were as they should be,

> everyone found "Dr. Mac" interested, helpful, and kind, full of cooperation and a willingness to try new ways; his frequent response to some new challenge was: "Well, why not?" But in the face of what his Scots Presbyterian psychological outlook conceived as wrong-doing, slovenliness, or careless human imperfection, he grew stern and flinty and became the advocate of moral and intellectual rectitude. As a student in his early classes at UCLA, I learned to toe the mark properly and exactingly, and later as a young faculty colleague it was clear to me that neither phase of his character had changed one iota (Spencer 1972, p. 430).

Interestingly, under McBride the UCLA department grew larger than the Berkeley department that overshadowed it. It came as no surprise that the distinctive characteristics of both their chairmen, Sauer and McBride, would clash. Yet students from Los Angeles would venture north to Berkeley to pursue graduate degrees; thus, their work at UCLA, while looked at with some raised eyebrows, was considered credible (Spencer 1979).

This was the situation in which Joseph Spencer found himself when he began graduate work under Sauer in 1929. He completed his field studies in the American Southwest in 1932, and then left for a position in China. Returning briefly in 1936 to defend his dissertation, his work in China continued until his appointment on the faculty at UCLA in 1940. He was called to serve with the OSS shortly after the beginning of the war (Spencer 1979). Spending more than two years in Southeast Asia, Spencer was promoted to Lt. Colonel, despite his candor and open criticism of the insensible policies developed behind the desk by Washington bureaucrats (Nelson 1985).

The war interrupted the department's efforts to establish a doctoral program. As McBride was engaged in work outside the university, Clifford Zierer, who had arrived in the department in 1925, the same year he was granted the Ph.D. at Chicago, took over the chairmanship. It was under Zierer that the graduate program was expanded to include the award of the Ph.D. in 1947 (Spencer 1979). Subsequently, two former students of McBride that received Ph.D.s under Sauer, Joseph Spencer and Henry Bruman, took on chairmanship duties as well.

Spencer continued to conduct studies in Asia; however, because of his work with the Nationalist Chinese government, he was not able to return to China. Thus he shifted his focus somewhat to Southeast Asia and conducted substantial research in the region. When his historical geography of the region was published, he received perhaps the greatest compliment possible, as it was from mentor and taskmaster, Carl Sauer: "Joe my boy, you've done (us) up proud by the first American regional geography that is really cultural geography. . . . When I'm asked in the future about the nature of geography, I'll refer them to this" (Nelson 1985, p. 599).

Attracted to academe because of scholarship in teaching and research, Spencer believed he was privileged to serve on a higher education institution's faculty. His sense of obligation compelled him to serve on committees across the institution, not because he had a strong desire to do so, but because he believed it was an essential duty in academic life.

> He was willing to serve because he fought any administrative usurpation of faculty rights. He never fell into the trap of feeling committee service was an accomplishment in itself; it was one way you paid your dues for the privilege of a scholarly life. Further, he felt an obligation to the larger community. Joe never turned down an invitation to speak and appeared with equal enthusiasm before both scientific groups and storefront gatherings (Nelson 1985, p. 598).

Perhaps the best evidence of Spencer's transformational leadership behaviors came from his colleague, Howard Nelson (1985), who reflected on his friend's excellent attributes. "Necessary . . . is the courage to take large chances and the wit to see opportunities as they appear. It helps if you have the insights to see what others miss and determination in the face of adversity. Above all it takes the willingness of a lifetime of plain hard work in your chosen profession. Spencer was one of those rare souls who was blessed with all the needed qualities" (p. 601). He did indeed attract students. Marvin Mikesell commented that he was among the students that Spencer recruited: "I drifted into geography, I suppose, through anthropology and international relations, history, things of that sort. I discovered there was a department at UCLA; I had never heard geography was a university subject. I talked to Joe Spencer, found it very attractive and so signed on. I've been aboard ever since" (Dow 1971).

Henry Bruman also appeared to be blessed with the qualities needed to serve as a leader in a higher education institution. He was indeed a meticulous scholar, having completed his Ph.D. under Sauer in 1940, and like so many other geographers, he served in the OSS during the war. Bruman was a Latin Americanist, and his dissertation on alcoholic drinks among the indigenous peoples of Mexico was published as a book and considered a fine piece of work on the subject. Bruman was also a great philanthropist, and because of sound real estate investments, he died a wealthy man. He left all to endowed professorships, scholarships, and departments at UCLA (Thrower 2005).

Looking backward, with a doctoral program in existence for only ten years, it took a special team of academic geographers to work together to build a department of excellence that was ranked sixth nationally in 1957. The fact that two of the distinguished figures that led the department were former students of the program lends credibility to the notion that alumni act as important stakeholders and view things through the long term. Interestingly, despite the role played by James Chamberlain, it was the women that led the department early on that appeared to establish a legacy of transformational leadership, a type of leadership behavior that characterized the department chairs for forty years. Though George McBride could legitimately be called the "Father of UCLA geography," much is owed to Myrta McClellan for laying the groundwork for the program's foundation.

NOTE

1. More than likely, it was Isaiah Bowman, Huntington's former colleague at Yale, and then McBride's supervisor as the director of the prestigious American Geographical Society in New York, who served as somewhat of a mentor.

SEVENTEEN

The Legacy of Carl Sauer

Transformational Leadership at Berkeley

When Ruliff Holway convinced Carl Sauer to leave Ann Arbor for the chairmanship at Berkeley, which carried the reputation of being the first independent geography department in the country, it was already the second time Holway had made Sauer an offer. In 1923, when Sauer finally arrived, Holway was then retired, and the new hire created quite an impression early on. Frederick Teggert, a founder of the university's Department of Social Institutions, and a confidante of Isaiah Bowman, wrote to the director of the American Geographical Society expressing his dismay over the brash manner in which Sauer presented himself. Sauer's reputation by that time had evidently already become known in geographic circles. Bowman responded: "The Sauer episode could have been forecast with mathematical accuracy" (Dunbar 1981, p. 69).

Carl Sauer was born in Warrenton, Missouri, the son of a college professor whom he deeply admired. At the turn of the century, the town exhibited a cosmopolitan atmosphere, as it played host to a Methodist college known not only for its emphasis on the liberal arts and sciences, but also a devoted, albeit poorly paid, faculty. Teutonic ancestral origins were important in the county, as a fifth of the county's population claimed German ancestry, and an even higher number were able to speak the language. Affiliated with the college was an orphanage the community had established after the Civil War to provide a home for the children of deceased veterans, and the higher education institution made a commitment to provide free education to the orphans. Here, Carl Sauer attended school and his mother frequently worked as a volunteer, demonstrating evidence of a familial commitment to serving those in need. An emphasis on German culture and literature, in addition to an extended family trip to the ancestral homeland, which included study in German schools, enhanced his education (Kenzer 1985). Hence the setting of Sauer's early years creates a picture of a social landscape that shaped a mental model marked by academic rigor, intellectual inquiry, and a sense of service to humankind.[1]

After receiving the baccalaureate, Sauer ventured north to the Chicago area, and initially studied geology in Northwestern University's graduate program, subsequently finding it unsatisfactory and moving on to the University of Chicago after only a year. At Chicago, he studied under Rollin Salisbury, the eminent geomorphologist who was not inclined to accept the simplistic notion of environmental determinism prevalent at the time. Platt (1957) recalled: "[Salisbury's] personal interest beyond physiography was attracted more to man's influence on nature than to nature's influence on man. For instance, in the field, I recall his looking at hummocky marshland and pointing out the influence of cattle in producing this landform—but never anything of what the land had done to the cattle or the people" (p. 276).

Sauer also was influenced considerably by Henry Chandler Cowles, the notable ecologist who pioneered research in plant succession on Lake Michigan's shores. Leighly (1976) commented that it was from this association that he "acquired ideas he later used frequently, both in relation to plants and, in a metaphorical sense, when writing of human geography" (p. 338). After a year of teaching at the normal school in Salem, Massachusetts, he completed his doctoral dissertation on his native Missouri's Ozarks, and then was appointed to the faculty at the University of Michigan.

Rising quickly to the rank of professor, his popularity (and perhaps his arrogance as well) raised the ire of the geology and geography department chair, W.H. Hobbs, who appears not to have made any attempt to encourage Sauer to remain in Ann Arbor when the latter decided to move on to California (Leighly 1979). It was in Berkeley where Sauer (1925), conversant in German and French, digested European geographical scholarship and penned "The Morphology of Landscape," instilling a *raison d'être* among numerous students who would pursue cultural geography, a new field of geographical study in America, and one introduced by Sauer himself.

It was this piece of work that sparked a new paradigm in American geographical thought devoted to the historical study of regions by accentuating culture as a principal agent for transforming natural landscapes (Martin 1985). Upon arrival to Berkeley, Sauer commenced field studies in Mexico with graduate students, a number of whom recollected their journeys and the inspiration drawn from their mentor (Kenzer 1987). His new graduate program was a "one-man show." Sauer did not seek out faculty from elsewhere in the United States, for he did not consider American graduate programs to be reputable centers of geographic education. He commented about the many land-use mapping dissertations completed in Midwestern departments: "There is nothing significant to be found out in such a thing. You map this year's crops until your area is mapped, and you are through. There will be none of that here" (Spencer 1979, p. 48).[2]

Some professors from Germany arrived in the 1920s to take visiting appointments, among them: Oskar Schmieder, Gottfried Pfeifer, Albrecht Penck, and Wolfgang Panzer (Leighly 1979). Notable figures in geographical scholarship themselves, they took on active roles in their own research as well as the education of undergraduate and graduate students. Schmieder, for example, conducted research in Mexico's Baja California with Sauer and a group of graduate students, publishing the earliest scholarship on the Guadalupe Valley (Schmieder 1928).[3]

When reduced enrollments occurred in the summer, Sauer invited some colleagues in the Midwest to Berkeley for instruction, among them: Wellington Jones, Kenneth McMurry, Stanley Dodge, and Preston James. Additionally, the Department of Geology invited William Morris Davis in 1925 to teach a seminar for a semester. Leighly (1979) reminisced: "It turned out to be, however, not a seminar as I understood the term, but a continuous monologue by Davis, in which he expounded ideas long familiar from his *Geographical Essays*, to which he had nothing new to add. The only relief was when he occasionally took chalk in hand and exhibited on the blackboard his admirable skill in drawing. The greatest of all was the greatest disappointment" (p. 7). Hal Raup also commented on his disappointment when asked, "What kind of a person was William M. Davis?"

> A rather stiff neck as you can imagine. I was so bold as to challenge one of his pet theories. Perhaps it was a mistake, but I went up after class one day and said: "If this Golden River really formed the Golden Gate shouldn't there be some terrace evidence on both sides of the Golden Gate at the present time?" He went right through the roof and that was about the last question I asked of him. You didn't challenge his theories, not in public, you didn't (Dow 1973).

A number of geography students took geology courses as a required minor field of study, while others studied anthropology. Further, Sauer required his students to have some literacy in German because much of the work done in geography that he embraced was published in German, and Leighly (and later Kesseli) introduced views that disputed Davis in seminars and publications. Leighly commented in an interview,

> [I]deas later than Davis' didn't get introduced into geography in this country and yet there were many new ideas. For example, when I was a student I was asked to read Davis' *Geographical Essays* at all three different institutions, at three different levels of study, and by the time I got through that in Berkeley at the graduate level I was thoroughly

> tired of Davis. So I rather disturbed the seminar (which was in the Geology Department) in which we were asked to read Davis by bringing in Hettner's *Die Morphologie der Erdoberfläche,* which was distinctly anti-Davisian. My instructor had never heard of Hettner and here I brought in these heretical ideas of Hettner and read them to the seminar. I'm afraid that the instructor was a little bit disturbed by that, but after all, if people had begun to read Hettner around 1920, perhaps something of the spell of Davis might have been broken (Dow 1970).

Moreover, Albrecht Penck, one of Davis's intellectual adversaries happened to be visiting as a professor at the same time Davis happened to be lecturing and Raup stated,

> It was very entertaining, because here I was working under both Davis and Penck at the same time and to get them into an argument with each other about their theories was a fascinating thing. They would call each other names and then laugh. It was a very entertaining experience. . . . To get them to joshing each other was very entertaining for a student. . . .I think it was without doubt [planned by Sauer as a confrontation of ideas]. I'm sure he had this in mind (Dow 1973).

In any event, by and large, professorial duties were assumed by Sauer's early cadre of devoted students. John Leighly, only six years younger than his mentor, was tasked with teaching both cartography and climatology. At the time, no textbooks on the subject existed; however, Sauer had faith in the teaching associate that had followed him from Ann Arbor as well as R.J. Russell, a geology doctoral student. Leighly (1979) remarked,

> I doubt that anyone of my generation or later was ever given responsibility for upper-division courses with so little formal preparation as I had. The instruction in geography I had received did not remotely approach what is normally required for an undergraduate "major." Such indifference toward the undergraduate backgrounds of graduate students was habitual with Sauer: few who earned the doctorate under him had taken their bachelors' degrees in geography. He trusted the students' native intelligence and ingenuity to overcome such obstacles as they encountered in their reading and in their original investigations, and in my case in my teaching (p. 6).

Geography graduate students also were required to take courses in other departments. Alfred Kroeber allowed them to take his seminars in anthropology, while ancillary coursework could also be taken in geology, soils, and botany. A rare female graduate student was able to take an advanced course in philosophy, and some students took seminars in ethnography. The small department made up for its limited size by capitalizing on the eclectic nature of the field and encouraging capable students with a passion for excellence and intellectual inquiry to study under sister disciplines' masters (Leighly 1979). Fred Kniffen discussed this in an interview:

> I had Sauer on the one hand, who gave me a tremendous orientation in geography and talked about the great figures in the field. He was the one who introduced to me the concept of "landscape;" it's been a central theme that I've never neglected.
>
> Oscar Schmieder was in Berkeley at the time, and there were other people, visitors who came through, who greatly influenced me. In anthropology there was Alfred Kroeber, Robert Lowie and Erland Nordenskiold, the Swedish anthropologist, who had been working in Latin America. From all these people I knew that somewhere there was my field.
>
> But the main thing that finally turned me to geography was the fact that in anthropology there was too little consideration of the physical landscape. After all, I liked the physical earth and felt that it was essential to my interest; so geography seemed the solution to filling my desires and aspirations in research (Dow 1976).

Moreover, the field studies most students early on conducted in Mexico united them, as did their single purpose in deliberating over geographical inquiry with regard to a particular problem, which was a marked contrast to the encyclopedic and dull regional studies that were the norm in American universities. Graduate students were relatively free to pursue their own interests in other departments, and nearly all were funded with assistantships. In Leighly's (1979) final analysis, he commented: "Best of all, we acquired a sense of participating in the worldwide and perennial enterprise of intellectual discovery, of which the university, the department, the course of the instruction, and even the single lecture, are pars, however minute. We caught a glimpse of Isaac Newton's great unexplored 'ocean of truth,' on the shore of which we might, with effort and some luck, pick up our own small handfuls of smooth pebbles" (p. 9).

This was the program that Sauer designed. With an unmasked arrogance in the pursuit of knowledge and disregard, as well as disdain, for the research American geographers were conducting elsewhere, his students, who also happened to be his staff, were driven to excel. Graduate students knew they were part of something special.[4] Students' reminiscences elsewhere do not often reveal condescending attitudes toward other departments; but at Berkeley, geography in other programs was scorned (Leighly 1979; Spencer 1979).

When Leighly was awarded the first doctorate in 1927, he became an assistant professor in a department whose small size was more of an asset than a liability. His colleagues, Thornthwaite, Kniffen, Dicken, and Meigs, were awarded Ph.D.s soon afterward and went on to other institutions, having distinguished careers in their own right. With Russell's award of the doctorate in geology, the need for an associate in geomorphology sparked the hiring of John Kesseli in 1932, who was appointed to the rank of assistant professor six years later when he was awarded the Ph.D. Jan Broek, a Dutch geographer who had completed his doctorate at Utrecht, but was funded from a fellowship to work out of the Berkeley geography department during its completion, was appointed to the faculty for several years before the Second World War. For much of the 1930s and 1940s, the department numbered no more than four full-time faculty, and by the end of the Second World War, twenty Ph.D.s had been awarded, placing Berkeley fifth on the list of total dissertations accepted in American geography (Rugg 1981). [5]

The superciliousness of Sauer's students was perhaps justified. Clark produced more doctorates than any other institution, with eighty-two dissertations completed by war's end; but its program was regarded as mediocre. Wisconsin, which was third on the list, also was not well regarded, and fourth-place Michigan's program left much to be desired as well. Only second-ranked Chicago, with an exceptionally strong faculty, and one that was relatively large, produced a steady stream of geography Ph.D.s. With the exception of Washington, another mediocre doctoral program that had awarded seven Ph.D.s by 1946, Berkeley was the only geography doctoral program in existence west of the 100th meridian, lending it to geographical isolation as well as the seemingly mandated insulation from the discipline in the larger departments of the eastern half of the country (Rugg 1981).

Moreover, the dissertations at Berkeley represented a wide variety of subjects and regions. The first ten awarded reflect this diversity: "A study in urban morphology: the towns of Mälardalen in Sweden" (Leighly 1927), "Louisville, Kentucky: a study in urban geography" (Thornthwaite 1930), "The Colorado Delta, Mexico" (Kniffen 1930), "The Big Barrens: a study in the Kentucky Karst" (Dicken 1931), "Historical geography of the Dominican Mission frontier, Baja, California" (Meigs 1932). "The Historical Geography of Northwestern Chihuahua" (Brand 1933), "The Pennsylvania Dutch at the Forks of the Delaware, Northampton County, Pennsylvania" (Raup 1935), "The Middle Virgin River, Utah; a Study in Culture Growth and Change" (Spencer 1936), "Cultural Geography of the Prairies of Southwest Louisiana" (Post 1937), and "Pleistocene Glaciation in the Valleys between Lundy Canyon and Rock Creek, Eastern slope of the Sierra Nevada, California" (Kesseli 1938). Sauer's role as a mentor guiding his doctoral students in pursuit of knowledge in a wide array of geographical locations and phenomena reflects not only the eclectic nature of the discipline, but also Sauer's intellect. He was truly a "Renaissance Man" in the middle of the twentieth century, conversant with the intellectual elite about things ranging from soils and plants to politics and culture. With the mainstream of society, he was knowledgeable about sports and other goings on. He appeared to be in his element, according to his students, when he would be talking to farmers, whether in the Santa Clara Valley of California or Valle de Guadalupe in Mexico, about their crops, their families, and their communities. Talented in generating dialogue with people from all walks of life, Sauer gathered information, stored it in the back of his mind, deliberated over it, synthesized it, and ripened it for discussion in a future seminar or a publication (Denevan and Mathewson 2009; Kenzer 1987; Leighly 1976; 1979; Mathewson and Kenzer 2003). Parsons (1996) reflected on his mentor and said, "Sauer was above all a geographer, a designation in which he insistently reveled. His stage was really the world and through all of human time" (p. 384).

Henry Bruman (1996), one of Sauer's earlier students commented: "Mr. Sauer was a wonderful man, a fine role model, and an inspiration to several generations of doctoral students at the University of California, Berkeley. We students tried to emulate him in various ways, consciously and unconsciously, but none of us could attain the intellectual heights he reached" (p. 370). Wilbur Zelinsky, another doctoral student at Berkeley in the 1950s, was asked about his recollections of Sauer and he responded:

> That he was a remarkable pedagogue. In the classroom he had a facility I've never seen any one, almost no one else, have. He spoke in perfect prose. You could take down his remarks delivered without any notes and transcribe them and print the results. He also had a trick I had never seen duplicated. He would end the class hour, when the bell rang, with a period and at the next meeting would pick up with the next sentence without even clearing his throat. In private he was, and I speak from my personal recollections (I'm not sure about others), we had a warm relationship. He was caring and helpful, strict, high standards and all that. I think the most remarkable thing about him was that he was an infallible judge of character. He could see things in people that nobody else could see.... He had a knack of discouraging people that he felt did not meet his standards and he made certain they went elsewhere. . . . He had a number of ideas, many of which turned out to be correct on matters involving historical and cultural geography. Another key element in his character was an extreme tolerance of a wide diversity of viewpoints, ideological, political, and intellectual. He would receive and talk to people we would regard as wacks, people who had wild ideas; he would give them a hearing. He would encourage all kinds of venturous research into obscure and difficult areas of investigation; some which paid off. No, he was not dogmatic, he was open-minded. He had strong ideas. For example, he had no patience with anybody, who thought that man had not been in America a very long time. I won't mention his intellectual opponents, but he thought ancient man had been in America for much longer than the then conventional wisdom would have had it. I think he was correct. . . . And he had other ideas, which then seemed wild, which I think have been

> borne out. A man of brilliant insights. By the way, although personally rather conservative in his views, he did publish toward the end of his life an article, which has now been picked up by the women's movement on the origins of agriculture and the importance of women (Dow 1984).

In essence, the impact of Sauer was pervasive, and the West became his domain in which he established an academic dynasty. In the classroom, Sauer was known as a captivating professor in his rambling lectures on North America and South America, in addition to thought provoking seminars on a variety of other topics pertaining to his thoughts and research (Denevan 1987). Yet, the influence of Sauer's leadership was felt in a number of arenas even beyond the small cadre of faculty and students that assembled with him at the University of California. He engaged in substantive research and publication that continues to be widely digested in the United States and abroad today, including a strong argument for the origins of agriculture in Southeast Asia, as opposed to the traditionally accepted Mesopotamia (Sauer 1952). Mikesell (1987) noted that Sauer "may still rank among the most quoted of American geographers" (p. 146). He served on a number of Association of American Geographers committees in addition to being elected its president and honorary president; additional service included his role as an adviser for the Office of Naval Research, Guggenheim and Rockefeller foundations, and several editorial boards.

Sauer tirelessly campaigned for research funding and wrote strong letters of recommendation for his students seeking employment. "If a full-length biography of Sauer is ever completed it will have to include a substantial chapter devoted to his role as a politician and entrepreneur" (Mikesell 1987, p. 146). His students held the utmost respect for him and because of his influence, the departments at the University of Oregon and Texas A&M were nicknamed "Little Berkeley" and "Berkeley on the Brazos," respectively. It's little wonder that distinguished professorships have been named in Sauer's honor at the University of Wisconsin and Louisiana State University.

Carl Sauer indeed articulated a vision that not only his students and colleagues embraced and pursued, but his academic grandchildren and great-grandchildren did as well. Certainly, his charisma and intolerance for mediocrity inspired his carefully selected entourage to question the norm, think outside the box, and strive for excellence.[6] It is difficult to imagine a more influential figure, because not only did he shape the course of people's lives by encouraging them to pursue geographic paths of inquiry, he essentially established a field within the discipline, cultural geography, which is practiced by adherents of subsequent generations today and has no intention of departing anytime soon, regardless of technology and a shift in paradigms.

Sauer shared his ideas freely. When he left Michigan, he provided Preston James, his replacement, with his notes for teaching classes in which James had no experience (Dow 1970; 1984). He gave Finch and Trewartha the outline that established the content of physical geography and resulted in their highly popular textbook (Leighly 1976; DeVivo 1988). And he advocated tirelessly for the discipline and the department both on and off the Berkeley campus, longing for practicing the geographer's craft, commenting in a 1937 letter to Isaiah Bowman: "I am at work as a geographer after a heavy siege of campus and off-campus activities. I still prefer trying to be a geographer" (Leighly 1976, p. 341). His passion for the field seemingly went unabated. Parsons (1976) remarked that Sauer worked on most weekends in his office in an attempt to satisfy that unquenched thirst for knowledge.[7] It is a paradox that his work was not considered in geography's mainstream, and this somewhat disturbed him, for he lamented to Preston James in a letter: "I don't think that the American geographer knows the stuff I've been doing" (Leighly 1976, p. 344).

Parsons (1976) reflected on his mentor in a tribute, commenting, "Even in his lifetime he achieved the status of a legend" (p. 83). His honors were many; included among them were honorary doctorates from Syracuse, Glasgow, Heidelburg, and the University of California. "Sauer had an undeserved reputation for being difficult to approach, 'the Great God West of the Sierra,' as geographers who did not know him occasionally whispered, apparently in a combination of grudging respect and disparagement. In reality he was a warm and friendly man, gentle, understanding, and easy to talk with, who served in the role of *pater familias* to many" (Parsons 1976, p. 89).

Martin (1987) wrote that "since his passing perhaps more sustained attention has been given to this luminary than any other in our history" (p. xvi). In essence, Carl Sauer was truly a transformational leader, arguably the greatest one geography in America has seen.[8]

NOTES

1. A history of service appears to be a relatively common attribute of transformational leaders in American geography. Sauer, McBride, Cressey, and White rank among them. This is quite a contrast to those transactional leaders marked by personal histories characterized by entitlement.

2. On the surface, it would appear that the condescending attitude Sauer directed toward geography doctoral programs in the Midwest was uncalled for; however, if one today were to examine a smattering of the dissertations produced before the Second World War (outside of Berkeley), Sauer's views can be rather easily understood.

3. Incidentally, I have used Schmieder's work in my own research on the geography of viticulture and enology, as this is Mexico's premier wine producing region today. The point being made here is that the work of Sauer, his colleagues, and his

students have withstood the test of time. The same cannot be said for the scholarship produced by many, if not most, academic geographers from the historic past.

4. When Ed Hammond related to me that even though he was encouraged to return to Wisconsin, where he received the M.A. in 1940, with the award of the prestigious Whitbeck fellowship, he chose to remain at Berkeley because he recognized that it was far superior, and the challenges imposed by Sauer and his small group of subordinates demanded excellence. After the Second World War, Hammond completed an exemplary dissertation under the guidance of John Kesseli on the geomorphology of Baja California.

5. In my conversation with Yi-Fu Tuan, John Leighly and John Kesseli each were marked by strong personalities themselves. Sauer was viewed as being akin to a dominating father figure.

6. I have known a number of his thirty-seven former students. In conversation, each held Sauer in high regard and accorded him deep reverence.

7. I recall talking with a former librarian of Berkeley's Bancroft Library in 1989, who commented to me that even in retirement in the late 1960s and early 1970s, Sauer regularly could be found in the library, reading and conducting research in historical geography.

8. The best way I can emphasize this point is to quote a colleague who met Carl Sauer making a rare appearance at a national conference: "When Sauer walked into the room, the temperature changed." Such was the impact of the man nearly two decades after his retirement. In essence, no academic geographer influenced more students of the discipline than Carl Sauer. This was corroborated in a conversation with Ron Abler, for he indicated to me that one of the many projects he never finished was oriented around a postcard survey of geographers asking about their dissertation supervisors and who had been most influential in their intellectual development. The influence of Carl Sauer was very obvious; concerning his "intellectual children and grandchildren," there were "a very large number of them."

EIGHTEEN

Leaders in a Paradigm of Eclectic Pluralism

The era marked by a paradigm of eclectic pluralism[1] characterized a period in which several programs established during the first half of the century experienced relative stability. Assurances gained by a renewed emphasis on the achievement of higher education in the United States, which was sparked by the G.I. Bill at the end of the Second World War, fueled a time of relative prosperity for most institutions, as society's demand for college-educated public citizens increased considerably. The conflict in Vietnam facilitated the mass annual migration commencing each fall from the homes of parents to university campuses, as draft deferments often were awarded to those young adults seeking higher education in particular fields, such as teacher education devoted to the elementary, middle school, and secondary grades. Often frustrated and unsuited for teaching in the public schools, new schoolteachers frequently found the master's degree to be the ticket to becoming college faculty in former normal schools that were now state colleges with swollen enrollments.[2] These institutions continued to attract new Ph.D.s as well, for the number of geography doctorates awarded could not keep up with the demand generated by open assistant professor positions in American higher education. Subsequently, second-tier, third-tier institutions, and sometimes programs in institutions that should not have belonged in any tier at all, offered the doctorate in geography to keep up with the demand. And though much interchange occurred between and among top-tier institutions with regard to graduates moving on from one place to the next, often Ph.D.s from "Top 10" programs found themselves in departments of lower status. Though the demand remained high nationwide for Ph.D. geographers, the faculty slots to fill in the highest ranked programs were limited.

Thus, keen competition in scholarship generated prosperity in academic geography during the 1960s and 1970s, and that scholarship was measured in publication records distinguished by those manuscripts that were included in refereed journals. Whereas an earlier paradigm was marked by the foundation of programs of academic distinction, which often, but not always, rested on the transformational leadership attributes of one important chairman, this new paradigm was marked by an absence of significant formal leaders as department heads. Rare was the department chair that held on to the reins for a decade or score or more of years, as did Barrows, Sauer, Miller, Hudson, Finch, D.H. Davis, and Cressey. Far more frequently, chairs rotated in already established programs for one or two elected terms, as the position tended to be one that had evolved from substantial authority and decision-making to the administrative management of one's peers. It was often perceived as an undesirable and time-consuming role to play, as status was frequently accorded to more informal leadership roles, with especial reference to scholarship. Thus, the position of chair, as a formal leader in geography programs, largely denuded; upper echelon programs sought to maintain their status quo, and to do that, it was seemingly more important to concentrate on generating scholarship in massive volume. Trees would be toppled, forests would be cleared, and paper would be made for the pages of scholarly journals that included articles written, some useful and some not, to justify the academic geographer's existence in the university.

At times a league of leaders would emerge that would dictate the trajectory of a program. For example, as Butzer has astutely noted, at Wisconsin, the oligarchy of Clark, Robinson, and Hartshorne determined salaries and raises behind the façade of a democratically administered department (Gould and Pitts 2002). At other times, leading scholars might take turns in administering the department, but still determine, by and large the program's direction. Such was the case at Minnesota, for example, where Borchert set the tone in the 1950s for a rotating chairmanship that was followed by Mather.

Thus, during this period of academic geography's history, a certain blasé fell over several departments that had achieved high standing in the national rankings. Satisfied with the status quo that resulted from reputable legacies of an earlier paradigm, some programs flourished as a consequence of past prestige while others floundered. In previous years, though competition between and among graduate programs in geography existed, a limited number offered the doctorate, so achievement of some prominence was not necessarily an arduous task. The commentaries provided by Trewartha (1978; 1979) and Prunty (1979), for example, point to the absence of rigor demanded of graduate students at Wisconsin and Clark. Yet, in the 1960s, those Ph.D. programs that could claim origins during the immediate post-World War II period or before could afford to rest on their laurels.

As the century continued on into the 1970s and 1980s, they became more frequently characterized by collections of individuals, rather than teams working toward a common purpose, for in the absence of distinguished figures marked by transformational leadership behaviors in formal leadership roles, there appeared to be little incentive to move individual faculty beyond each individual's goal. The vision that had been articulated and realized by transformational leaders of an earlier generation was later enjoyed by those who had been able to reap the benefits of the foundations their predecessors had laid down; however, it is questionable as to whether or not many academic descendants of stalwart figures such as G. Donald Hudson, Cressey, or even Sauer, carried on the leadership traditions of their mentors.

Rather than devotion to a particular department or institution, progressively more important became the national society devoted to the geographer's craft: the *Association of American Geographers*. At one time an exclusive organization that had been so bold as to deny membership to a standing U.S. president who had earned the Ph.D. in history and political science two decades before its founding, the AAG had developed into a more egalitarian group, providing forums, both regionally and nationally, for an increased interconnection that facilitated friendships, alliances, and recognition among its many members. Individual geographers increasingly became more inclined to pursue stature by wrestling with the political affairs of the organization rather than the politics of their own institutions, though the latter occurred as well, and certainly publication records played important roles. Though it had occurred from time to time in an earlier era, it became increasingly common for individual faculty to be receptive to seductive offers of distinguished professorships at peer institutions, or departments on the rise, rather than remain at the same institution for the duration of their careers. For example, in the 1970s, Gilbert White left Chicago for Colorado, and in the decade that followed, Karl Butzer went from Chicago to Texas, and Yi-Fu Tuan moved from Minnesota to Wisconsin, sparking in each destination institution an elevation in status and image. The reality of the times fostered a higher level of allegiance to scholarship and the geographic discipline, as opposed to devotion to the academic department.[3]

At Berkeley, this was not the case. After Sauer retired from the chairmanship in 1956, his student and colleague John Leighly, who had been with him since his days in Michigan, assumed the chairmanship. Leighly had led a remarkable life and contributed much in geographical scholarship, despite being in Sauer's shadow throughout most of his career. He had served in France during World War I, gained substantial knowledge in the normal school in Mount Pleasant, and acquired a solid education in Ann Arbor. Dismissed from his assistantship in geology by the chairman for placing his feet on his desk while talking to a student and taking too many courses on the geography side of the department, Leighly left the

University of Michigan and went on to Berkeley with Sauer, completing his Ph.D. in 1927. His interests were eclectic, though he accomplished much in climatology, urban geography, and cartography, and he advised many students. Leighly was known as an especially empathetic and caring professor who was marked by a high level of integrity (Miller 1988); however, he was not inclined to engage in endless streams of paperwork and did not enjoy his tenure in the chairmanship. He was more disposed to serve as a mentor. Miller commented: "Many graduates of the department from the 1930s into the 1980s, speak of him as a friend who considered students colleagues, not disciples and who often socialized with them.[4] Decades later they recall his generosity in some matter and his painstaking help in their lonesome struggle to enter the world of independent scholarship" (Miller 1988, p. 348). After his death, his colleagues would write:

> He disliked the drudgery of administration and committee work. He spoke openly of the irksomeness of his years as departmental chairman immediately prior to his retirement. He has written how he "dreaded seeing those kraft-paper envelopes used for intra-campus communications that come down from higher levels of the administrative hierarchy." He was probably too conscientious to be a chairman. He answered every letter of inquiry from prospective students in great detail, but as their enrollment typically required unavailable funding, such effort went for naught.
>
> He was a kind and courteous colleague, uncompromising, direct, soft spoken, and wise. He did not unbend easily, being by nature shy and quite formal, a true gentleman scholar. He was frugal both in his personal life and as an administrator. He had no desire to see the department grow and, never forgetting that the university's funds came from the taxpayers, he was hesitant about increasing departmental expenditures when making out the year's budget in his role as chairman. He set a standard of scholarship and intellectual integrity that contributed significantly to the strong sense of identity among Berkeley geographers that still persists. He represented as no one else could the continuity that has characterized the department over the past six decades (Parsons et al. 1986, p. 176).[5]

Following Leighly, another student of Sauer, James Parsons, took over as chair. Perhaps just as much as Leighly, he showed great kindness to students and colleagues. Also a remarkable scholar and exemplary teacher, he showed numerous students the way to the discipline and facilitated the success of many, while also inspiring colleagues to excel. Certainly, a transformational leader, as was his mentor, it was said, "more than anyone else in Berkeley geography, even Carl Sauer, Jim was the spirit that held us all together" (Wallach 1998, p. 323). No more seamless a transition in leadership could have occurred from the time of Sauer's retirement until the end of Parsons's first term as chair ten years later; Unfortunately, during that term an external review criticized the department and "recommended new hires that would break away from what some people saw as an exhausted orthodoxy" (Wallach 1998, p. 323). The cultural geography that had characterized Berkeley for decades as a unique program unlike the mainstream of geography became viewed as largely irrelevant, and it would be replaced by a "'critical human geography' and a concern with social theory that put it very much in the mainstream of contemporary social science and literary studies." By the time Parsons retired, after a second term as chair in the 1970s, he was indeed disturbed by the change in the departmental paradigm, commenting on the "collapse of the cultural tradition at Berkeley," "a diversion to the political left," "intellectual arrogance," "ruins the department" (Wallach 1998, p. 323). Wallach's last words on his mentor must be noted: "If he had been less kind, Berkeley Geography might have lasted a bit longer at Berkeley. Yet, if he had been less kind, a whole generation of geographers at Berkeley would never have known Berkeley Geography" (Wallach 1998, p. 324).

In essence, this "Berkeley Geography" that had persisted for forty-three years under the chairmanship of Sauer, Leighly, and Parsons was forced to undergo an uncomfortable transformation. Diversifying the faculty included the addition of stellar scholars seeking to divorce themselves from the department's traditional past. Over the course of the next forty-three years, seven chairs attempted to administer an increasingly disruptive and disagreeable collection of academics that reflected a very different type of geography than was practiced in the Sauer era. [6]

Similarly, at UCLA, Sauer's disciples stood fast in the 1960s, only to lose ground with the addition of faculty from outside the established realms of the discipline. Henry Bruman and Joseph Spencer had been leaders of the "Old Guard," supporting the Berkeley tradition and opposing the newer spatial science. In fact, Spencer was editor of the *Annals* during much of the 1960s, and it was reported that he asserted, "I will never publish a paper with a mathematical equation in it."[7] As the years wore on, a split was apparent between those representing the human and physical sides of the discipline. Further, as the faculty diversified, philosophical differences were exacerbated, which created a marked impact. For example, adding to the faculty in urban geography was William Clark who arrived in 1970.[8] Nicholas Entrikin joined the faculty as a social geographer in 1975, and in 1981 came Allen Scott, an economic geographer, who commented to me that the three worked in concert as a "New Guard" to transform what was perhaps "the last major geography department in the United States into an up-to-date department of geography. We pushed, and pushed, and pushed, and there were a lot of retreats." There were also a lot of tensions as well, and no strong leader emerged to take over the chairmanship in the last two decades of the twentieth century. Yet, it may not have been a department in which a chair could do much more than serve in a quasi-administrative capacity, for the true leadership may have been held informally by individuals and alliances that carefully navigated the academic landscape by identifying objectives, wielding power, and achieving goals.[9]

In any event, elsewhere in academe, students of Sauer's Berkeley School had already made their mark. At Wisconsin for example, Andrew Clark, continued in his role as mentor to a strong following of historical geographers until 1975. William Denevan, a student of Parsons, served with distinction there as well until his retirement at the end of the century.[10] David Ward, one of Clark's early students (and thus an intellectual grandchild of Sauer), returned to Wisconsin in 1966 as a member of the faculty. He subsequently served in a number of administrative leadership positions, eventually directing the institution as its chancellor.[11] In 1983, another Berkeley alumnus arrived; Yi-Fu Tuan, a luminary in humanistic geography took the J.K. Wright Professorship twenty years after he had been rejected for an academic appointment as an assistant professor at Madison. The department by this time was large, vibrant, and relatively young; not yet fifty-three years of age, Tuan was among the most senior members of the department. As the chairmanship continued to rotate and the oligarchs had by now retired, a sense of community developed that fostered academic prosperity.[12]

Elsewhere in the Midwest were positive changes as well. At Minnesota, for example, in the 1950s, enrollments were booming. A number of instructors were hired to accommodate student demand, and among them was Fred Lukermann, a veteran himself and a Ph.D. student with field experience in Turkey. With the departure of his advisor, John Weaver, Lukermann gave up on the Ph.D.[13] Borchert, serving as chair between 1956 and 1961, had gained substantial influence in the university, and when the academic administration sought to terminate the instructors not holding the doctorate, the department "cashed in all its chips" and got Lukermann advanced to "associate professor and tenure." By the early 1960s he became the department chair, in 1966 he became the associate dean of Social Science, and in 1967 he was appointed the assistant vice president. Lukermann was regarded as a brilliant intellectual—"a polymath"—and along with Borchert, the two were considered strong leaders on campus. Moreover, they energized good students.[14] They were not alone, however, in capably leading one of the top geography departments in the country.[15]

In the 1960s and 1970s, a number of factors appeared to be correlated with the success of doctoral programs in geography, and one of them pertained to the acquisition of funds through various revenue streams, one of which was usually undergraduates; if a department did not address undergraduates, the university could not support it. In the early 1960s at Clark University, not only were undergraduate enrollments in geography low, but the director of the School of Geography was rather introverted and not inclined to build relationships that would fuel success. Ray Murphy, the editor of the journal *Economic Geography*, had become the school's director in 1962; but internal dissent about the direction in which the program was heading led to his resignation in 1965. Perhaps fifteen to twenty graduate students were enrolled, but the program was small and it appeared to be in deep trouble; there was even talk about closing it down altogether.

When Saul Cohen accepted the invitation to take over as director in 1965, he found a program with six members and six factions; but he thought he could bring consensus to Clark. Three faculty retired, and those that remained were inspired by his vision of revitalizing the geography program, despite realizing that the institution had no money to contribute to the cause.[16] Yet Cohen drew from his experience as the Executive Director of the Association of American Geographers and his network in Washington, DC, to gain funding to build a program. Acquiring funds made available through the National Science Foundation, National Defense Education Act, and U.S. Office of Education, while also facilitating the endeavors of his colleagues, stellar programs at both the undergraduate and graduate levels were formed. For example, he realized that Martyn Bowden, Robert Kates, and William Koelsch were not being used in the most effective way, so he enabled them to concentrate on their individual strengths. Bowden was a wonderful lecturer, but he didn't write much; he was able to concentrate on his teaching and undergraduate enrollments soared. Kates was very bright and good with graduate students and Koelsch was a "research techniques person par excellence."

Cohen also was not bashful about allying with other disciplines. As Clark was known for its highly regarded psychology program, collaboration fostered developments in environmental perception. Moreover, faculty holding credentials outside geography were welcome at Clark, and he refused to allow his own philosophical views dictate hiring decisions. Cohen related to me, "We helped open the field of Geography to other disciplines. I hired a political scientist and a psychologist. . . . I was not a radical geographer, but there was room for radical geography at Clark." In essence, "getting people to think beyond their normal borders," worked, and his leadership behaviors proved to be successful. Subsequently, becoming dean of the Graduate School, it was said about him that "his visionary qualities always allowed him to see the larger picture and never to lose sight of the ultimate goals. He led by example and commanded respect. He took on a lot of work himself and expected the same of those around him, so that most of his colleagues found themselves wanting to live up to his expectations" (Waterman 2002, p. 564).

In 1978, Cohen left Clark to become president of Queens College, CUNY; but it is important to note that the success of his efforts lived on in

Worcester after his departure. In 1982, Clark Geography for the first time in its history entered the "Top 10" in national rankings. Before the end of the century, the faculty numbered more than twenty. In essence, it was through the transformational leadership of Saul Cohen that Clark Geography prospered.[17]

Three hundred miles from Clark was Syracuse, also enjoying success as one of academic geography's stellar programs. Preston James had been serving as the chair of the geography department from the time George Cressey stepped down in 1950 well into the 1960s, and it must be noted that the James perceived by some to be an unbending "stuffed shirt" at AAG meetings was not the same James who presided over the department at Syracuse. James Newman[18] related to me, "He tried to help everybody that came into the department. . . . He didn't have a vindictive part of him. . . . He sold the discipline. He sold the department." James also was known as "a master of collegiality," and scheduled monthly social gatherings at his home. "He wanted all faculty to participate on an equal basis," and these gatherings became an important part of the department.[19] It was here that members of the department shared geographical ideas and items on their own individual research agendas.

When James left Syracuse for retirement in Hawaii, however, no longer did the faculty meet regularly. Donald Meinig, a brilliant and widely revered historical geographer, briefly took over the chairmanship; but he had no interest in administrative duties.[20] Shortly thereafter, Robert Jenson took the chairmanship and served in that role for seventeen years. During that time, physical geography was deemphasized,[21] and a shift in emphasis to political and urban geography took place. Research in the historical geography of the Americas continued to be engaged in, as were other aspects of (mostly) human geography; but the chairmanship of Jenson was a sharp contrast to that of James. His long chairmanship appeared to enable him to make contacts in the administration that facilitated his rise to becoming the dean of the College of Arts and Sciences as well as to the dean of the Graduate School. Interestingly, three faculty were hired while he was chair with no searches or departmental discussions. This sort of practice did little to foster collegiality, and the sense of community within the department declined. Subsequent to Jenson's term, John Mercer was appointed chair, and although the department continued to rank highly, the camaraderie that James built in an earlier generation seemed to be lost.

Meanwhile, at Penn State, although the department's academic standing was on the rise, internal dissent threatened the program's integrity. In 1963, Allan Rodgers took over as head when Will Miller became an assistant dean, and at the time, the program included six additional faculty: George Deasey, Phyllis Griess, Peirce Lewis, Paul Simkins, Fred Wemstedt, and Wilbur Zelinsky. Peter Gould, a pioneer in quantitative methods who had briefly served on the faculty at Syracuse, was then hired, and by the end of the 1960s, John Adams and Ron Abler, young economic geographers with Ph.D.s from Minnesota, had joined the faculty, and with Tony Williams, the four "young Turks spearheaded a new graduate course in geographical analysis with a strong quantitative basis" (Haggett 2003, p. 927). David Ley would reflect (Aitken and Valentine 2006), "Their own momentum was enhanced by the stimulating presence of David Harvey and Julian Wolpert as visiting professors, by the appointment of Roger Downs as a superb instructor, and by the visits of Gunnar Olsson and Les King, other members of the 'Michigan Inter-University Community of Mathematical Geographers', who showed up periodically for energetic conversation at the departmental watering hole on College Avenue. Peter Gould was the animator of these events and an extraordinary intellectual presence" (p. 180).[22] Yet, Gould also proved to be a divisive loner, courting his own students, in a department that already had faculty not getting along.[23] Neither he nor Lewis ever served as department heads, and when Zelinsky reluctantly took over in the early 1970s, his sometimes poor judgment and budgeting skills, which left much to be desired, created problems. Ron Abler subsequently became the head of the program and administered it soundly over the course of time, before taking leadership positions with the National Science Foundation and Association of American Geographers. By 1995, Roger Downs was head of a program that included twenty faculty. Despite the conflict and absence of esprit, the department had risen to rank first among American geography graduate programs by the National Academy of Sciences.

Thus, several highly ranked geography programs grew to list many faculty on their rosters, and some were able to offer nearly all things geographic to all people interested, which exemplified the new paradigm characterized by eclectic pluralism that prevailed. Certainly, department size played a significant part in each department's continued prosperity, for it was rare for significant leaders to materialize, as did the transformational leaders in the earlier generation that established each department's success. Faculty in the same department often would find dialogue easier with colleagues in related disciplines in other programs, for in these top-tier programs, and even in mediocre ones, it was not unusual for specialists in fluvial geomorphology, social theory, gender and inequality, climatology, pollen analysis, historical geography, and cultural geography to share space under the same roof.

At other leading institutions, such as the University of Washington and Ohio State University, rather than offering diverse curricula, the impact of the quantitative revolution compelled the departments to focus on numeracy and spatial analysis. Something as esoteric as cultural geography would not be considered acceptable at either place; at Washington after Hudson resigned from the chairmanship in 1963, significant changes occurred in the university administration. Administrative changes resulted in the department being tasked with justifying its existence as John Sherman, the leader of one of three significant centers of cartographic instruction in the country,[24] took on the role of chair for a decade. Though he did not demonstrate the same caliber of leadership as his predecessor, for Hudson was a tough act to follow, to maintain the advancements the department had made during the tumultuous times of social turbulence that followed the Kennedy assassination was an accomplishment in itself. The 1960s were a time of campus unrest across the country, and in Seattle it was no different. These difficulties only compounded the tragedy that Sherman felt in the middle of his tenure as chair, when in 1967 his son had been killed in Vietnam. Richard Morrill succeeded Sherman and following him were Morgan Thomas and William Beyers. Though the university never matched the administrative support it had extended during the 1950s when Hudson was chair, the department continued to achieve prominence in national rankings (Velikonja 1994).

Climbing steadily in the rankings throughout this period was Ohio State University, which owed much of its elevated status to the transformational leadership of Ned Taaffe, a Ph.D. from Chicago who spent much of the 1950s at Northwestern University.

> In 1962 the Department of Geography at Ohio State was looking for a new chairman to replace Guy Harold Smith who was retiring. Several prominent geographers were approached about the position, all refused. It was widely known that the department's reputation had been in decline for several years and that there was a real possibility it might be eliminated. When Ned was approached about the possibility of becoming the new chairman, he was less than enthusiastic. Neither he nor his family wanted to leave the Chicago area. However after the dean of the College of Business, James McCoy, met Ned, he was convinced Ned was the right person to build a research department and restore its reputation. He made an offer Ned could not refuse, including five new faculty positions and the financial support to offer attractive salaries. In June 1963, Ned took over the chairmanship at Ohio State (Gauthier 2002, p. 576).

Taaffe accomplished what other transformational leaders had done elsewhere in an earlier era with support from a committed administration; he continued to produce sound scholarship, even with his administrative duties, and he articulated a vision for the program, which captivated many new people he chose to be members of the Ohio State geography faculty. He succeeded in his efforts by bringing to the university a group of committed and vibrant young geographers willing to excel; among them, Leslie King, Reginald Golledge, and Lawrence Brown were but a few that worked with Taaffe to establish Ohio State as a center of geographical excellence.

> His leadership soon transformed the department into one that was recognized as among the best in the nation. He hired a young faculty that complemented his own vision of geography. As chairman, he was fair and reasonable in his demands and always supportive of initiatives and endeavors of his younger faculty. Many would themselves become leaders in the quantitative transformation of the discipline during the 1960s and 1970s. In time, several would receive various honorary recognitions from the AAG. A few would even be elected to the association's presidency, a position in which Ned served in 1971–1972 (Gauthier 2002, pp. 576–577).

Although the department had placed eighth nationally in 1957, when the 1964 Cartter rankings appeared, Ohio State plummeted to seventeenth place. After five years as chair, Taaffe was able to guide the department back into the top tier as the seventh best program in the country (Warf and Webster 1980). He retired in 1992; but before his death in 2001, Taaffe would see the department he had spearheaded move from the verge of extinction to fifth in national rankings by the end of the century (Goldberger et al. 1995). Ned Taaffe was indeed a compassionate and caring intellectual, and one of the most important transformational leaders that epitomized the paradigm of eclectic pluralism, for not only was he

> the architect of the strong Ohio State University graduate and undergraduate programs, [he assumed] the role as the "bridge between spatial analysis and traditional geography." . . . While many of the new "quantitative geographers" were zealots who discarded geographic tradition, Ned sought to place the new perspectives within the intellectual heritage of the field. . . . The program he developed at Ohio State produced geographers who not only were articulate in analytical methods but knew how to use them within the conceptual framework of geography (Gauthier 2002, pp. 577–580).[25]

Behind Ohio State and Washington, the University of Iowa continued to plod along near the top of the second tier, also placing strong emphasis on the new spatially analytical traditions (Warf and Webster 1980). Although its administrative leadership did not appear to be exceptional, what it accomplished with limited support from an oftentimes uncooperative administration was substantial.[26] Its small number of faculty, in comparison to its peer departments in other institutions, could be proud of what they had achieved (Lindberg 2006).

Geography at Johns Hopkins also persisted near the top of the second tier in the three decades after the Lattimore affair (Warf and Webster 1980). The presence of human geography subsided considerably following the departure of Lattimore in 1963 for the University of Leeds in Britain to assume the directorship of a new program in Chinese and Mongolian Studies. Within a few years, his nemesis, George Carter, left Hopkins as persona non grata for a distinguished professorship at Texas A&M University. Shortly thereafter, Cambridge-educated David Harvey left Britain to serve on the faculty as a geographer of some distinction, and he became known as the world's most cited geographer, especially for his works in theory, social justice, and critical geography. Except for a six-year period in which he served as the Halford Mackinder Professor of Geography at Oxford, Harvey spent most of his career at Hopkins, where geographers were, perhaps not so remarkably, scarce. He commented that his only colleague on the human side of the department for most of his tenure was educated as a planner (Gould and Pitts 2002). In any event, Harvey's leadership lied in his scholarship, which was exemplary and revolutionary. The administrative leadership of the department rested in Gordon Wolman, and here one must raise the question about whether or not Harvey would have accomplished all he has done without the support of the transformational leader who was his chair.[27]

Wolman was a highly regarded fluvial geomorphologist with strong ties to Hopkins; his father had distinguished himself as an exemplary professor on the sanitary engineering faculty for many years; Wolman would follow in his father's path, not only in terms of scholarship, but also in administrative leadership, for he had the makings of a transformational leader. Wolman's friendship and regard for Lattimore facilitated stability during what might have been otherwise perilous times. This was exemplified by his invitation for Lattimore to return to the department and lecture on matters associated with the human geographical dimensions of the field.[28] Moreover, after Harvey had left for Oxford and sought to return to Hopkins, it was Wolman that was able to arrange for his reappointment on the faculty, despite some opposition in the university. Wolman was "willing to go to the mat for something he believed in. He would really fight hard for principles. Everybody trusted him."[29]

Yet, Wolman's passion was in his own work, and he was able to articulate a vision that inspired many by merging geography and environmental engineering programs. Quickly, environmental engineering dominated; however, at Hopkins it was a natural match. Wolman reminisced, "I went from behavior of natural rivers and how they formed to how they behaved in the environment, which led to water quality and then to a variety of resource policy issues."[30] This epitomized the department's trajectory during the 1970s and 1980s, and although it became far removed from the emphasis on cultural geography and global affairs that had distinguished it somewhat during World War II and its aftermath, the narrowly focused specialization enabled the department to maintain a high level of credibility, despite its absence from the top tier.

Not all departments that carried stellar reputations in the 1960s were able to survive the budget cuts heavy-handed administrators laid down. Though not ranked as top tier programs in geography, closures at Penn in 1963, Stanford in 1964, and Yale in 1967 came in the wake of the Harvard disaster and precipitated departmental demise at some of the leading universities in the United States (Murphy 2007). The first significant geography program to fall was at the University of Michigan. There a collection of geographers had succeeded in keeping the program in the top tier, and in 1969, ranked behind Chicago as the second best department in the country (Warf and Webster 1980); however, the Board of Regents voted unanimously to close the department. Following Kenneth C. McMurry's more than thirty years as chair, came Charles Moler Davis (1956–1966), L.A. Peter Gosling (1966–1969 and 1971–1974), Melvin Marcus (1969–1971), Donald Deskins (1974–1979), and lastly, John Nystuen (1979–1982). Waldo Tobler recalled:

> Geography at Michigan was in good shape, but they had made some poor decisions. Several of the faculty had affiliations with area centers or other units and were thus de facto only part-time members of the department. Three senior geographers—Gunnar Olsson, Melvin Marcus, and I—left at the same time, and the department was able to replace these positions only with nontenured assistant professors, a mistake that weakened the department. Few of the faculty served on campuswide committees, another weak point. From my service on the Dean's Promotion and Tenure Committee, I was aware that a historian on that committee was antagonistic toward geography, arguing that Harvard did not have a geography department, so Michigan did not need one. All these factors, combined with a financial crunch in the state of Michigan, led to the closing of the department (Gould and Pitts 2002, p. 314).[31]

At the 1981 annual meeting of the East Lakes Division of the Association of American Geographers, George Kish, a senior member of the department, delivered an invited address, and he commented on what had happened:

> At 10:00 on January 27, 1981, I went to meet a class. At 11:00, I was met by my chairman, who, at that time, had just returned from the office of the dean to be informed that discontinuance proceedings were established, instituted against the department. We found out after the news, the radio and the television heard it. I think the best reaction was from the son of our first chairman, John McMurry, teaching at Waterloo, Ontario who called me that night and said he was out in the field with students. He had his radio in his car. He heard it. And he barely avoided the ditch. It was incredible.[32]

Four years later, Northwestern University folded its department, as did the only remaining geography graduate program in the Ivy League, Columbia.[33] Chicago remained a small bastion among the top private universities in the country, as geography, albeit in a wounded state, continued its presence until 1987 (Murphy 2007).

The legacy of success that had been built by giants such as Salisbury, Barrows, Colby, and Platt, was greeted by Gilbert White in 1956, who resigned his position as president at Haverford College in order to take the chairmanship at his alma mater in the Windy City. Truly a transformational leader, White, a devoted Quaker who had not only conducted humanitarian relief work for the Friends Service Committee, he had developed a solid reputation in geographical scholarship as well as higher education leadership. He completed the doctorate in 1942 and, among other things, engaged in work for President Roosevelt (Murphy 2008). He recounted in an interview:

> Pearl Harbor came and I had registered as a conscientious objector (a Quaker) and I couldn't very well stay in the Executive Office of the President. The Director of the Budget said: "Well, if the Draft Board asks me I won't give you permission to leave, but if they'll let you go you can go." So I volunteered to do Quaker Relief Work in Vichy France. The Board happily gave me permission (which they gave to very few CO's) to leave the country at that time. In France I worked in concentration camps, with the children in schools, and I spent a year detained in Germany, came back and was responsible for administering Relief in China and India from the Philadelphia Office. All that together accounted for about four years of volunteer service during the war. That led on to my going to a Quaker College, Haverford College, which seemed a logical next step for somebody concerned about peace and education. I stayed at Haverford for almost ten years. At Haverford I did just about all I felt I was capable of doing. I increased the endowment considerably, reduced the size of the student body, brought on a lot of good, very lively faculty members and new programs. By that time I felt I ought to get back to my first love of geography. Very fortunately, Chicago was willing to take me back; I had had an appointment when the war broke out to go back to Chicago. They took me back in 1956 in the same old department, which had changed very much in the interim.
>
> When I went as a student to Chicago the Department was organized on the basis of specialties and areas. By the 1950's (although Bob Platt had been very influential in his functional approach to regional problems) the Department no longer felt it had to cover every region of the world. Nor did it feel that it had a large obligation to provide instruction in regional geography as such; it was moving in the other direction. New appointments were made, such as Berry and Mikesell. There were appointments of people who had specialties that were more of a systematic sort. We were, in effect, demonstrating the kind of training that Ed Ackerman had been talking about - the shift from regional to systematic organization (Dow 1984).

White returned to the Chicago campus shortly after Ackerman decided to leave academe to devote himself full-time to examining water resources issues with the non-profit organization, Resources for the Future (Harris 1974). Interestingly, White had also conducted substantial research in water resources. "By the time White left the University of Chicago in

1969, he was one of the most influential voices in natural resource management in the United States (Murphy 2008, p. 447).

Other chairs of the department over the course of the next thirty years included Chauncy Harris, Marvin Mikesell, Brian Berry, William Pattison, Norton Ginsburg, and Michael Conzen, and some took on administrative leadership positions elsewhere in the university as well. For example, Ginsburg served as Assistant Dean and Associate Dean of the University's Division of Social Sciences of the undergraduate division of the University for several years, and he also contributed to scholarship in Asian studies. He took over the chairmanship shortly after the departure of Brian Berry.[34]

After being awarded the Ph.D. at Washington in 1958, Berry arrived at the University of Chicago as a new assistant professor, just twenty-four years of age.[35] A leader in quantitative geography, he captivated many seeking to apply statistics in human geographical study. Developing a reputation as a brilliant scholar, by the age of thirty he had risen to the rank of full professor.[36] Berry's academic role was complemented by his contributions in applied geography, which found him serving on a mission to India with USAID and consulting with the World Bank on a number of development projects in Asia, as well as meeting with the Nixon White House to address domestic policy issues. Developing a network that had been fostered by an effort to move beyond the halls of an academic environment seemingly characterized more by satisfaction with simply creating awareness of problems plaguing society rather than trying to solve them, he was able to contribute to the department's prosperity. With the departure of White, Berry remained as the only geographer on the faculty that brought in considerable grant money to fund graduate students.[37] When he left in 1976, hoping to reestablish geography at Harvard, the department at Chicago found itself in a much weakened state. It folded in 1984.[38]

Karl Butzer, a faculty member on the staff, saw the writing on the wall, the apathy was noticeable: "Geography had lost three of eight full professors in 1968–69, without even attempting to replace them. After 1971, endowments and scholarships dried up, as did the influx of top-caliber students. But I had learned at Wisconsin not to try to change the academic establishment" (Gould and Pitts 2002, p. 79).Yet the academic administration expressed unwillingness to support the department because of internal strife, and that discord was fueled by Norton Ginsburg, whose hobby of disparaging Brian Berry extended beyond the department to not only the academic leadership of the university, but to the Ford Foundation in India, which had sought his counsel in metropolitan planning.[39] Berry (2006) commented, "The dean told me that since university budgets were getting tighter, he would not provide new support for the department, or even replace existing faculty, until internal differences were resolved" (p. 155).[40]

Chauncy Harris, who at one time had served as the dean of the Division of Social Sciences and Director of the Center for International Studies, became especially disturbed by the department's decline. Marvin Mikesell (2004) commented, "Harris was perceived generally to have been a judicious dean who acted wisely within the constraints of a never-adequate budget" (p. 988).[41] One might think that with the networking that would have occurred, the department was on solid territory; however, the upper level administrators that had supported the department and were part of Harris' network of allies had retired. Mikesell (2004) commented,

> It is safe to say that Harris's greatest and perhaps only regret about his University of Chicago career was that he could not do more for the Department of Geography. When serious problems became evident in 1984, he had reached mandatory retirement age and experienced a precipitous decline in his influence. Moreover, most of the important people he knew well and respected his judgment, notably President Edward Levi and Provost and later President John Wilson, had also retired (p. 988).

Marvin Mikesell, a student of Sauer who had been at Chicago since the late 1950s, led the department at various times until the program's termination and eventual devolution to a committee on geographical studies. The department had done well professionally in national and international arenas in terms of scholarship and politics; however, it had obviously done poorly *vis-a-vis* political affairs at the university. The chairmanship became a revolving door, and quite obviously a number of the department's members failed to do what needed to be done to argue for their existence; as opposed to the transformational leadership attributes exemplified by White and his predecessors, leadership at the departmental level apparently transitioned to one seemingly increasingly marked by laissez faire behaviors.

One might argue that it was perhaps the shift in the 1950s from a collection of regional specialists to a smattering of thematic geographers (albeit solid authorities in their respective fields) to blame. Regional geography was not dead, just somewhat redefined, and the exemplary Soviet specialists that continued to be produced at Chicago until Harris's retirement presented evidence of its value, for significant contributions to the discipline were made; however, his colleagues did not necessarily attach themselves strongly to particular world regions, and in a small department with a rich tradition, devoted to a discipline that is often considered esoteric by outsiders, it is the study of places—of regions—that often fosters its credibility. Having said all this, the fall of Chicago appears not to have had anything to do with the nature of geography and the chosen pursuits of its practitioners, but more to do with the personalities that clashed and apathy in promoting the program among leaders in the academic administration, as well as the loss of funds to support graduate students after the departure of White and Berry. The dean of Social Sciences terminated the program.[42]

Moreover, after the president of Haverford College returned to Chicago to chair the department, it appears that unlike the previous generation, faculty were not cultivated to take on formal leadership roles—or perhaps they were disinclined to do so. The elimination of geography at the University of Chicago presents a sad chapter in the history of American geography, for it was here that the first graduate program in the discipline was founded, and it would not reappear for some time, if ever.

Alexander Murphy (2007) commented that the termination of geography programs at Harvard, Penn, Yale, Michigan, Columbia, Northwestern, and Chicago "could be seen as the result of actions of unsympathetic administrators or of internal problems (lack of attention to undergraduate education, feuds among faculty, etc.). But focusing on such explanations directs attention away from the larger issue: many in the U.S. academic elite had come to view geography as a dispensable subject to institutions of higher learning" (p. 125). Though this might indeed have been the case, and it might remain so, one of the principal problems has been, and continues to be related to weaknesses in leadership. And this is a matter that still needs to be addressed.

As the 1980s drew to a close, academic geographers recognized that marked changes had occurred, and specialists in various fields found more in common with other disciplines than with geography. Practitioners in palynology, political economy, and feminist geographical theory increased in prevalence, and the concepts of space, place, and landscape became blurry visions of the past. The spatial perspective, central to geographic inquiry, became considerably weakened in the approach taken by many that professed to be practitioners of geography. Those that longed to hold on to the traditional modes of research were compelled to justify their existence. In a new era that embraced numeracy and technology, it was difficult for the geographer from an earlier generation who went out into the field and got his or her boots muddy to find common ground with those that explored the world gazing at computer screens seated in ergonomic chairs. And Berkeley wasn't Berkeley any more. A major transformation was about to take hold.

NOTES

1. *Eclectic pluralism* was coined by Martin (1985).
2. As many former schoolteachers that had sought academic degrees in education to escape the draft grew disinterested in public school teaching, they found homes as faculty members in state colleges after receiving master's degrees. Yet many were not suitable in academe, for their scholarship was lacking considerably in comparison to

those actual academics that had engaged in doctoral study. Feeling out of place (and often inferior), those with less rigorous academic backgrounds often became envious of those with greater intellectual talents. Thus, it was not unusual for many of them, from Connecticut to California, to devote their efforts to union activities and criticisms of the true scholars within faculty ranks. Consequently, departments were often fractured between actual academics and the academic wannabes.

3. Yi-Fu Tuan discussed with me his move from Minnesota to Wisconsin as a response to a "midlife crisis" where he had reached a plateau, although he had been very happy in the Twin Cities. Brian Berry (2006) indicated Gilbert White related to him that ten years were long enough in one institution to accomplish one's goals. Karl Butzer received a lucrative offer from a second–tier department wishing to elevate its standing in the rankings. The reasons for the migration of senior faculty are many; but endowed professorships and hefty salaries do make nice enticements. And those fortunate enough to be awarded in this manner are usually deserving of these incentives; however, it would be naïve to think that the institutions they have chosen to leave are not often adversely affected.

4. Yi-Fu Tuan commented in conversation that Leighly, although characterized by a strong personality and deeply intellectual, appeared embarrassed when he was tasked with examining the graduate student Tuan in German (as the intimidating Kesseli who was usually charged with these duties was on sabbatical). Leighly was fluent in German; but his view of graduate students was that they were colleagues, and his "extraordinary modesty" made testing a colleague of his uncomfortable for him.

5. In the last year of Leighly's chairmanship, he surprised his friends by marrying a graduate student thirty-six years his junior. Incidentally, it has been remarked to me by those that knew them that although there was a significant age gap, they were very much in love. On the first day of his formal retirement their son, Joseph, was born, and later came a daughter. A brilliant scholar and a benevolent soul, he continued to lend assistance to graduate students while he raised his own children during his retirement (Parsons et al.1986).

6. Since the mid-1980s, one would hear geographers comment and oftentimes lament again and again at national conferences and in conversation, "Berkeley isn't Berkeley anymore." The department had indeed changed significantly from the one that had been known for cultural geography. The intervening years between Parsons non-consecutive terms as chair saw Clarence Glacken, a scholar devoted to examining the history of ideas relating to man and nature, and David Hooson, a Soviet specialist, at the helm. Turmoil took place during Glacken's term as chair, and the stress led to (a nervous breakdown, heart attack and) his premature resignation. Subsequently, Hooson took the chairmanship and made efforts to diversify the department and move farther away from the cultural tradition. After Parsons' last term as chair ended in 1979, Allan Pred served in the role. With the exception of two brief periods (1980–1981 and 1985–1986), he served as chair until 1988 (Bernard Nietschmann and David Hooson filled acting roles as chair during his absences). A former student of Brian Berry's at the University of Chicago, Pred was highly regarded as a leading scholar in urban geography. Impressed with his delivery of a Carl Sauer Memorial lecture, the department hired David Stoddart, a notable physical geographer at Cambridge, to be the new chair; he served in that capacity until 1994, tendering his resignation after a number of faculty complained to the vice-chancellor. The chairmanship then fell to Richard Walker, who kept hold of it until 1999. Walker, a Hopkins alum, was known for his research in radical geography, economic development, and water resources (http://geography.berkeley.edu/about/history.php). In essence, the chairs in the last twenty years of the twentieth century appeared as mid-level administrators attempting to herd a dysfunctional band of scholars—not an unusual situation in academe. What might be considered unusual, however, is that despite the warring factions, philosophical differences, tenure battles won and lost, and bouts of alcoholism and

illness, which often plagued its chairs during the last forty-or-so years of the twentieth century, the department persisted.

7. Allen Scott, UCLA Emeritus Distinguished Research Professor, related this to me in conversation, and Brian Berry (2006) has discussed the *Annals* editor's refusal to accepting articles incorporating quantitative methods as well.

8. Howard Nelson had already been on the faculty for some time before Clark arrived, and his work was solid; however, he did not practice the quantitative geography that was embraced by others.

9. Despite the disharmony that existed within the institution's geographical community, a genuine interest in student success appeared to be a priority. William Koelsch, professor emeritus at Clark University, related to me that when David Angel, a 1989 UCLA Ph.D., was being considered for an assistant professor position at Clark, every member of the UCLA faculty was asked to comment on his suitability for the position. Surprisingly, despite their reputation for conflict at the time, every single response was a highly favorable endorsement of David Angel. Allen Scott told me that while a graduate student, Angel was very dynamic and enthusiastic about economic geography and was a de facto leader of sorts among the geography graduate students. "He meant a lot to the group of graduate students. A very sociable guy. A very pleasant kind of guy. Very personable." Scott recalled that he said at the time, "When he gets a job, he will become chair of the department." In 2010, David Angel was named president of Clark University.

10. Among the rotating chairmen, Denevan and the physical geographer, James Knox, were known as good departmental administrative leaders.

11. Yi-Fu Tuan related to me that in a number of academic leadership positions, David Ward was known as "an excellent administrator." Ward served as the chancellor of the University of Wisconsin for two non-consecutive terms: 1993–2001 and 2011–2013. He also served as president of the American Council on Education (2001–2008). A notable historical-urban geographer, he fostered the faculty's confidence in the upper echelons of the administration at Madison by being concerned with the needs of enhancing a scholarly guild rather than seeking to accentuate the typical management styles often practiced by those in higher education leadership positions.

12. According to David Ward, governing academic deans wanted department chairs to serve three-year terms, and this had actually been a major impetus for the department to embrace a protocol oriented around the notion of rotating department chairs.

13. Weaver and Broek developed a rift, and following Weaver's departure he took on a number of academic leadership appointments throughout the U.S., including the presidencies of the University of Wisconsin and the University of Missouri.

14. In separate conversations, John Adams, Ron Abler, and Yi-Fu Tuan have commented about Lukermann's strong intellect and solid leadership. Following his retirement, he was awarded the Ph.D.

15. Philip Porter and John Adams were among those considered excellent chairs before the turn of the century.

16. For example, Susan Hanson, who had impressed William Koelsch and Robert Kates at a number of regional conferences beforehand, arrived in 1981. By 1984, she was promoted to full professor and she took on the editorship of the *Annals of the Association of American Geographers*. By 1988, she became director of the geography program, and four years later took over as editor of *Economic Geography*. She was considered by her colleagues to be an effective chair, contributing much to the discipline, as well as Clark.

17. Indeed it was Wallace Atwood that introduced geography to Clark; but its credibility as a Ph.D. program was questionable. Moreover, his self-serving nature, questionable ethics, and anti-Semitic views (he refused to allow Jews to enter Clark's geography Ph.D. program) did little to facilitate trust, a necessary ingredient in leadership. Perhaps it is ironic that it was a Jewish geographer that arguably not only saved

the program from extinction, but also brought to Clark legitimate status and prosperity through his practice of transformational leadership.

18. Professor Emeritus, Syracuse University.

19. Through his fund-raising efforts and skill in relationship-building, James was able to secure funding for the distinguished Dell-Plain professorship.

20. In addition to Donald Meinig, David Sopher, who passed away in 1984, was influential at Syracuse.

21. The geomorphologist, Edwin Hammond, who had left Wisconsin to serve on the faculty at Syracuse, departed in 1970 to become head of the department at Tennessee. The biogeographer, David de Laubenfels, remained as the only physical geographer.

22. Before his premature death, Gould was recognized with the award of an Evan Pugh distinguished professorship, the highest honor given by the university to a Penn State faculty member.

23. Rodgers and Zelinsky didn't talk to each other, "though they did a good job of concealing it from students." And Gould sparked a conflict between physical and human geography. He asserted that physical geography was passé, and Peirce Lewis, "at heart a geomorphologist," of course resented it.

24. The other cartographic centers were at the University of Kansas, led by George Jenks, the University of Wisconsin, led by Arthur Robinson, and UCLA, led by Norman Thrower. Incidentally, Thrower had served as a cartographer for the British Army during the Second World War in South Asia; he subsequently studied cartography under both Erwin Raisz (at the University of Virginia) and Arthur Robinson (at Madison).

25. Taaffe had a background in meteorology, and he acknowledged the department's tradition in climatology since its inception; so he hired John Rayner to direct an atmospheric sciences program within the geography department, which was founded in 1971.

26. It is unfortunate that leaders in university administrations refuse to support solid academic departments, especially when it pertains to the award of tenure. I recall in the mid-1980s when Graham Tobin, as a young untenured assistant professor, received unanimous support from his colleagues in the geography department, as well as those on institutional committees, to be awarded tenure and promotion to associate professor. His record was exemplary and he showed much promise even then. The promotion was awarded; but the president refused to grant tenure. He went elsewhere and served as a faculty member as well as in academic administrative leadership positions, making news in 2013 by submitting a "principled resignation" from a top administrative post; this forced the president of the University of South Florida to modify the "draconian" budgetary measures she planned to implement (http://www.insidehighered.com/news/2013/08/16/u-south-florida-softens-financial-management-plan-after-administrator-resigns#sthash.tOyijhRw.dpbs). Indeed showing the markings of a transformational leader, one is compelled to wonder what might have been had Tobin remained at Iowa for the duration of his career.

27. David Harvey left Baltimore shortly after the turn of the century, later lamenting about his return to Baltimore: "Things just got too dismal for words in the department and the university, and when an attractive offer came to join the Anthropology Program at the Graduate Center of the City University of New York (with Neil Smith and Cindi Katz as colleagues), I jumped at the chance. It proved a shot in the arm and an opportunity to engage on an even broader canvas of thought and activity" (Aitken and Valentine 2006, p. 190).

28. For a period of time after the exoneration of Lattimore, Owen Lattimore and George Carter served on the faculty in the same department. John Boland indicated to me that acknowledging the tensions that existed between the two, Wolman held an annual party at his residence, with the expectation that all departmental faculty would attend. All did indeed attend, and Wolman was able to choreograph the setting year after year so that two different parties were simultaneously taking place within view

of each other; although guests might meander between the two, Carter and Lattimore would remain on their respective sides, and in this way relationships were facilitated from colleague to colleague across the department. It is also interesting that in his assignment of office spaces, Wolman placed individuals with dissimilar interests next to each other in order to further build relationships.

29. This was related to me by John Boland, who emphasized that Wolman was indeed a man of integrity that worked effectively for his department and the university, serving twice as provost. Another example of his belief in doing something he firmly believed in was his quest to get for Grace Brush, a research scientist who had been on the staff for many years, a faculty appointment; even though there was no opening, it was done.

30. http://eng.jhu.edu/wse/Reds_Wolman/memor.

31. In addition to Tobler and Olsson, three other geographers departed less than three years before the department was terminated: Bernard Nietschmann and Thomas Detwyler went elsewhere in academe, and Everette Bannister died in a car accident. Incidentally, Marcus had left in 1974.

32. Thanks are extended to Jeroen Wagendorp for extending to me a copy of an audio recording of George Kish's 1981 address. It should be noted that Kish also acknowledged the department had failed to adequately address the needs of undergraduates; nor did it generate publicity concerning its members' accomplishments or have adequate representation on university committees involved in decision-making that could (and did) influence its destiny.

33. Duane Marble related to me that when he joined the faculty at Northwestern, Taaffe was leaving for Ohio State as he arrived. Ed Espenshade was assigned the chairmanship, but his editorial obligations relating to *Goode's World Atlas* left little time for him to guide the department; however, he was surprised when he learned that Marble wanted to leave Northwestern, so he directed him to see the dean, in hopes that he could be convinced to remain in Evanston. When Marble entered the dean's office, he was asked: "You want to leave this place?" After Marble responded affirmatively, the dean replied, "I can't frankly see why anyone would want to stay here. I'm leaving myself."

34. Berry was succeeded by William Pattison, a profound leader in geographic education, who returned to the university's School of Education after serving a term as the Geography Department's chair.

35. Berry's (2006) memoir, *Nihil Sine Labore*, offers fascinating (and relatively unvarnished) accounts of his experiences in academe. A remarkable geographer, trained in historical geography by H. Clifford Darby in England before joining the graduate students in Seattle deeply immersed in quantitative methods, Berry was respected by academic geographers across the discipline throughout his career, which continues at the present age of eighty. This is noteworthy in that during this era of eclectic pluralism, dispute and disrespect for the discipline's members of different cloths was not uncommon. Editors of the leading geographical journals, for example, intentionally excluded articles from geographers of dissimilar persuasions. Yet Berry found acquaintanceships with not only the leaders in spatial analysis like Duane Marble and Waldo Tobler, he forged friendships with the ultra-traditionalist Richard Hartshorne and the humanistic geographer Yi-Fu Tuan. The cultural geographer, James Parsons, wanted to hire him at Berkeley and the distinguished climatologist, C. Warren Thornthwaite, sought to take him away from the human side of the discipline and convert him to physical geography. Among the stellar scholars of unlike fields mutual respect was accorded indeed. As Berry noted in a conversation with me, "People who are secure in their own skins have no problems communicating with people who have contrasting ideas. The problem arises with people who are not secure in their own skins." And unfortunately, it was often those beset by insecurities or afflicted with shabby scholarship that fueled disharmony and discontent.

36. Berry (2006) commented, "My rise through the ranks had been rapid—some thought too rapid. It was unseemly to reach the rank of full professor at the age of

thirty! At every opportunity, Norton Ginsburg made his reservations known to anyone in the university who would listen, but Chauncy Harris bless him (and unbeknownst me) ran interference" (p. 124).

37. Berry related to me that he and Gilbert White provided the departmental leadership and served as principal sources of funding for graduate students; in his (2006) memoir, he commented, "I wanted my students to learn that practical work should stretch clients, perhaps changing their world view, while key elements of applied work should be brought back into the university to challenge prevailing theoretical discourse. My colleagues on the other hand, argued that they should be allocated an 'equitable share' of the graduate students to work with them on 'more scholarly' pursuits. I was the obstacle, since most of the graduate students wanted to work with me" (p. 155). Harris was always a strong voice of support for Berry; but it appears that Ginsburg convinced an external review committee that Berry "dominated the department too excessively" (p. 154). Whether or not this contention rang true, the animosity directed at Berry from Ginsburg seemingly exacerbated tensions in the department as a whole, and this had been ongoing. A solid urban geographer, Harold Mayer had already departed for the University of Wisconsin at Milwaukee when he learned Berry was receiving a higher salary than himself. Further, Berry's negotiating strategy in departmental meetings, which was characterized by taking an extreme position on a certain issue with the intention of compromising to the desired objective, was certainly not a tactic deemed acceptable for the "thin-skinned." Yet it was Berry that brought in funding, inspired students, and contributed more than his fair share to the department's scholarly output. He certainly exhibited a number of transformational leadership behaviors, but his distaste for formal department teas, which appeared pretentious, and a healthy appetite for confrontational debate, which was encouraged at the university, lent to his reputation for being brash and abrasive. He probably could have done more in the cultivation of relationships among his colleagues, and perhaps he could have been an effective transformational leader as a department chairman if his colleagues in the department were cooperative; however, Norton Ginsburg, who had "butted heads" with Berry from the time he was a visiting professor at he University of Washington in 1957, was unwilling to allow his own ego to be overshadowed by Berry.

38. Berry (2006) acknowledged, "Many geographers blamed me. I had selfishly 'thought only of Brian Berry,' 'deserted Chicago,' and 'left geography'" (p. 157). It is true that it only takes one malcontent to create disharmony, fuel dissent, and bring ruin to an academic department. At Chicago, however, that person was not Brian Berry; it appears to have been Norton Ginsburg.

39. Berry (2006) remarked, "Norton Ginsburg had written Paul Ylvisaker at the Ford Foundation advising that they 'keep me in the back room' because I was 'too confrontational'" (pp. 103–104).

40. "Norton Ginsburg received the department chairmanship that he so coveted, but he became the captain of an ice-bound *Titanic*. As the collision neared, he conveniently passed the helm to Michael Conzen . . . but it was too late for him to change course, and it gave Norton someone to blame" (Berry 2006, p. 156).

41. Chauncy Harris was indeed highly regarded as a man of integrity. I talked with him at length nearly twenty years ago, and even then, in his early eighties, he did what he could to advance geography.

42. Interestingly, Gordon Clark, who had been promoted to associate professor and recommended by the institution's review committee for tenure, was refused tenure by the dean of Social Sciences, a sociologist. Professor Clark has had a distinguished academic career and is currently the director of the Smith School of Enterprise and the Environment at Oxford University.

NINETEEN

Simonett and the Santa Barbarians[1]

As I reflect upon my own entry into the discipline in the early 1980s, I recall that the eclectic pluralism, which legitimately characterized the prevalent paradigm, continued to be marked by philosophical conflicts that had been sparked during the *Quantitative Revolution* of the 1950s and 1960s. Membership in the Association of American Geographers was then half of the more than 10,000 affiliated with the society today, and certainly less than half of peer organizations, such as the Geological Society of America. Typical geography undergraduate degrees required a selection of courses drawn from a list of physical, cultural, and regional categories, in addition to cartography. Though some programs continued to require students to laboriously draft maps on Mylar plastic sheets with pen and ink, computer mapping applications had begun to replace the labor-intensive traditional cartography, and word-processing software was beginning to replace the electric typewriter. Students often were allowed the option of taking either a fourth semester of a foreign language or a course in statistics to complete the baccalaureate; their post-graduation employment, if successful in gaining a position in the field, was often in local and regional planning offices as assistant planners or cartographers. Several went on to work for the federal government in branches such as the Bureau of the Census, Department of Agriculture, Geological Survey, or the CIA, as their skills in designing, fabricating, and interpreting maps were considered highly useful in these agencies; but many students, with a breadth of knowledge, but no depth in any particular area, as was frequently the case due to the provincial nature of many geography programs, found themselves to be relatively unmarketable, and not much different than their peers that graduated in other social sciences disciplines.

The Landsat program had been launched during the previous decade; satellites in the earth's orbit could now record visual information and subsequently generate those images created to earth. The field of remote sensing had been spawned from cartography, for the lenses mounted on satellites capturing images of the earth's surface required individuals skilled in map design and geographical knowledge to interpret them. Hughes Santa Barbara Research Center had designed multi-spectral scanners—essentially, earth observing sensors—generating images of Yosemite's Half-Dome in 1969, the same year astronauts landed on the Moon. Also in that year, the University of California at Santa Barbara hired John "Jack" Estes as a faculty member in geography. Jack had just been awarded the Ph.D. at UCLA, completing a dissertation under the eminent cartographer, Norman Thrower, titled: *Applications of Multi-Image Correlation Remote Sensing Techniques*. And it is here that this stage of the story unfolds.

In 1973, a change in the administration at Santa Barbara resulted in the appointment of a chemistry professor, Bruce Rickborn, as the associate dean of the College of Letters and Science. Reaganomics had resulted in budgetary reductions, leaving the program with few opportunities for success; however, Rickborn had become convinced that the institution's geography program had been mismanaged. Members of the university academic senate sought to terminate the program, and Rickborn was tasked with playing an integral role in deciding whether or not to maintain it as a viable course of study. He appointed three academic geographers from outside the university to study the situation: James Parsons (Berkeley), Charles Bennett (UCLA), and Julian Wolpert (Princeton). Their several months of research provided Rickborn with a viable assessment of the discipline. He concluded, "Eighty-five percent of Geography departments were moribund, as exemplified by UC-Berkeley" (Clarke and Baumgart 2004, p. 99); however, he recognized that the work Estes was conducting in remote sensing was extremely valuable, as was the grant funding he was bringing in to the university. Rickborn decided that a program built around Estes could do much for the institution; so in what came to be known as the "night of the long knives," all geography faculty other than Estes were dismissed from their appointments at the university (Clarke and Baumgart 2004).

Needed was an effective leader to guide the program through its infancy and quickly into maturity, so the search for a new department chair was completed with great precision and care. Harm de Blij, known for his commentaries on *Good Morning America* and a noted textbook author with a wealth of experience in field studies the world over was interviewed and not considered further to lead the intended new program.[2] Reginald Golledge, a keen social geographer at Ohio State University, was much highly regarded in deliberations; however, the committee decided on David Simonett, an exemplary geographer with a solid record of research in remote sensing, and a wealth of experience in academe and private industry.

Simonett had been on the faculty at the University of Kansas for several years before returning to his native Australia to chair the geography department at the University of Sydney. He had sought to develop a program in geography oriented around the measurement and analysis of geographic phenomena *vis-à-vis* technological applications; however, his academic administration was disinclined to support this endeavor. He then took a position as director of Earth Satellite Corporation, and it was while there that he learned of the Santa Barbara opening; he viewed the founding chairmanship in California as an opportunity to develop the kind of program he had hoped to create in Australia.

Immediately upon being hired, he began a campaign to bring on board a highly qualified group of practicing academic geographers, skilled in technology and quantitative methods. He learned that Golledge had been the search committee's second choice, so he worked tirelessly to recruit him as a faculty member. Convincing the administration to provide recruiting funds, Simonett travelled to Australia, where Golledge was resting after a sabbatical in New Zealand.

> Pleasantly zoned on the beach, Golledge was extracted from reverie by someone kicking sand on him and barking the traditional Australian greeting of "Ow are ya, mate?" Turning his head to the left, the sunbather spotted a pair of white freckled legs, pants rolled up to the knees. Looking upwards, Golledge recognized the smiling face of David Simonett, who immediately asked, "Have you made up your mind yet? When are you coming to Santa Barbara?" Without waiting for an answer, Simonett crouched down and launched into the plans he had for the fledgling Geography Department at the University of California Santa Barbara, what Golledge was going to teach, specifics about salary and health benefits. He drew in the sand to illustrate and spoke as if it were a foregone conclusion Golledge would come (Baumgart 2004, p. 1).[3]

In any event, Simonett's tenacity paid off; when Golledge returned to Ohio State, he decided to make the move. Moreover, he phoned a close colleague at the University of Michigan, Waldo Tobler, who also had been approached by Simonett, and he resolved to make the move as well. Tobler had been part of the famous Garrison seminar at the University of Washington in the 1950s that was considered instrumental in geography experiencing a "quantitative revolution." A brilliant geographer, now he was on the cusp of another revolt in geographical thinking, as he joined like-minded colleagues in Santa Barbara.

With Simonett in the lead, new hires came on board dedicated to the research mission of the department. He fostered an atmosphere of encouragement and scholarship, sharing ideas, getting research grants for junior faculty, networking with administrators, and paying for work study students out of his own pocket to assist both students and faculty.[4] Jeffrey Dozier recalled,

> We thrived because of two of Simonett's characteristics. One, he talked to everybody. In other words, he kept everyone up to date on what everybody else was doing; he helped professors cross-pollinate. And two, he had the least contentious ego. He took genuine pleasure in other's achievements; he didn't need to take credit (Clarke and Baumgart 2004, p. 101).

He had a vision of creating a solid Ph.D. program in an institution that awarded only the geography baccalaureate when he arrived in 1975. A master's program was developed in 1977, and shortly thereafter came a Ph.D. The first Ph.D. was awarded in 1982, and in 1988, the geography department became part of the consortium that won the National Science Foundation grant for the establishment of the National Center for Geographic Information and Analysis (NCGIA), and the Santa Barbara department became one of the top geography departments in the country, if not the world, and it has served as a model for the establishment of similar programs elsewhere (Doctoral 2010; Goldberger 1995; Groop and Schaetzl 1997); for example, in the 1990s, the University of South Carolina subsequently initiated a doctoral program largely devoted to GIS, as did the University of Connecticut. Moreover, since then, GIS study has been included as an integral element in nearly all geography programs at all education levels, and here the role of Simonett and his Santa Barbara colleagues in influencing the course of geography has been pervasive, for with the establishment of an internationally reputable program nearly exclusively devoted to the application of geographical technologies to resolve issues and problems facing society and the environment, a new paradigm in geographic thought emerged that has revolutionized the content and course of teaching and research in higher education: *Geographic Information Science*.[5]

Certainly, the tenacity, vision and inspiration of David Simonett were driving forces in the department's success. He epitomized transformational leadership in many ways, for he noted not only the needs of the discipline (and society) to advance into the twenty-first century, he also recognized the opportunities made available by an academic administration willing to take risks; however, it was Estes's work that most likely convinced the institutional leadership to invest in geography. As a non-tenured assistant professor, Estes was already the third-highest source of external funding on a campus occupying a proruption on the California coast (Thrower 2005). He had essentially justified the existence of geography at a higher education institution marked by a number of faculty members and administrators that would have been happy to see it gone. The discipline had an advocate in the associate dean who recognized that a team of dedicated faculty inclined to pursue geography beyond the seemingly simplistic, "This tribe occupies this area, and this is what they grow and eat" (Clarke and Baumgart 2004, p. 99), could facilitate the university's mission and elevate its status. Wisdom and fortuitous circumstances led to Simonett's hiring to make things happen. And he did.[6]

For individuals like Golledge and Tobler, who had long been tenured faculty in "Top 10" geography departments,[7] a mid-career move to a brand new department without yet a graduate program required a form of risk-taking that most academics would not even consider. Yet, Simonett appeared to relay significant confidence in his hopes and plans for the future, and that confidence had already developed a record of diffusing wildly. Morain commented. "So contagious was his spirit and enthusiasm that by 1970 Dave had helped plant the seeds of remote sensing in dozens of departments, worldwide" (Morain 1991, p. 912).

At Santa Barbara, Simonett had the opportunity to select the members of the new department, and he chose his colleagues wisely, demonstrating his personal commitment to them in a number of ways. For example, he provided personal funds for his new colleagues to purchase homes in California's booming real estate market (Baumgart 2004). Moreover, instead of being concerned about potential challenges coming from the geographer that was not chosen to be the founding chair, but considered highly suitable nevertheless, Simonett pursued the hiring of Golledge with a level of determination that had him searching on a beach in Sydney for his prospective future colleague.[8] His hiring of Terry Smith also shows his character, discretion, and sense of leadership.

> Simonett published an ad in professional journals, seeking senior and distinguished professors who were firm believers in quantitative analysis. One of the respondents was Terence (Terry) Smith, from Rochester, New York. He wrote a 10-page letter. Simonett wrote back tersely: "Dear Dr. Smith, You are neither senior nor distinguished, but we think your work is interesting." Three or four months later, Smith's phone rang.
>
> Simonett: "Smith?"
> Smith: "Yes."
> Simonett: "This is Simonett."
> Smith: "Who?"
> Simonett explains who he is, then asks: "Are you still interested in a job here?"
> Smith: "Yes."
> Simonett: "When can you come out?"
> Smith: "In about a month or two."
> Simonett: "What about next week?"
> Smith: "How about in two weeks?"

> Smith made arrangements to fly out. As yet, he knew nothing about Santa Barbara. With amazing synchronicity, the airplane Smith took from Rochester to Buffalo, en route to UCSB, was named "City of Santa Barbara." Simonett, dressed in tennis shoes, met Smith at the airport and drove him directly downtown. He opened the car door at the 1400 block of State Street, saying, "Walk down to the pier. You really should look at this place and make your own mind up." Smith ambled down State Street to the wharf. Afterwards, he found his way to the Peppertree Inn, in the 3800 block of State Street, where he was to stay.
>
> During the ensuing visit, Smith interviewed and met people. When he was returning to Rochester, standing in line for the airplane out of Santa Barbara, an airport employee walked up to him and said he had a call on the courtesy phone. Smith left the line and picked up the phone.
>
> Simonett: "Smith?"
> Smith: "Yes."
> Simonett: "This is Simonett."
> Smith: "Yes."
> Simonett: "Would you like the job?"
> Smith: "Yes."
> Simonett: "You got it" (Baumgart 2004).

From screening and hiring faculty to numerous meetings devoted to developing a proposal for the NCGIA and seeing its establishment, to guiding graduate students seeking success, Simonett spearheaded the successful efforts of a stellar group of academic geographers from leadership positions ranging from department chair to dean. Certainly, his own scholarship not only lent much credibility to the program, but also to himself as a department chair, for he "left a legacy of more than one hundred published papers and several books and major technical reports for which he was the author or editor" (Smith 1991, p. 85). This record of achievement corroborated his authority, so when he spoke with candor, which he often did, he not only had the formal leadership position and the transformational leadership attributes that went with it, but also the scholarly research and publication record that validated his knowledge and views. Morain commented,

> Dave was a kindly but impatient sole *[sic]* who had to 'grow' on you. He would not suffer mediocrity or condone procrastination. Such people were 'ditherers of the first magnitude,' or suffered 'a congenital case of the slows.' If you didn't understand his true motive to stimulate you into action, you were doomed. He exuded a sense of—if something could be conceived, it could be done; so why hadn't it already happened? He was a ruthlessly honest critic who could not only grasp the essence of situations, but articulate the arguments in crystal clear terms that any observant person around the table saw were being met with nods of approval. On the other hand, you could tell when you'd said

> something that he hadn't considered. He'd get a quizzical look on his face, become very reflective or, sometimes, just walk away leaving you with the sense that you'd really spotlighted your stupidity. Then, sometime later, he'd raise the question again adding several illuminating thoughts that left you wondering if your idea was, after all, pretty good.
>
> His honesty actually inspired the otherwise unworthy to reach for higher goals. Mostly, it worked! (Morain 2006, pp. 764–765)

In any event, after seven years as chair, Simonett was asked to take on the position of Graduate Studies dean; Golledge then assumed the chairmanship, and he reflected on the mentorship Simonett had provided.

> With Simonett's guidance, I learned about university politics. Within the University of California system, the academic senate is very powerful. Simonett identified a number of key committees and made sure that our department was represented on each of them. Very quickly I learned the power structure of the campus and began finding where all the skeletons were buried. This knowledge was to stand in good stead when I later became department chair (Gould and Pitts 2002, p. 141).

Golledge continued to engage in solid research and publication as the department's chair, and it was under his leadership, with Simonett as the dean of Graduate Studies, that the ideas for the NCGIA were cultivated, which led to its eventual establishment in collaboration with Buffalo and Maine. During his chairmanship, the department experienced substantial growth; "as an 'Above Scale' Professor, Golledge played a key leadership role in the emergence of the UCSB Department as a top-ranking Geography Department in the US" (Stimson 2009, p. 338). Unfortunately, his tenure as chair was short-lived, for he was beset by blindness; Richard Church took over the chairmanship and oversaw the launching of the NCGIA and the expansion of the department. Michael Goodchild, a highly regarded specialist in GIS at the University of Western Ontario had been added to the department; he played a key role in the planning and implementation of the NCGIA. Goodchild became chair in 1998, and three years later the department was marked as the most significant grant-receiving program on the Santa Barbara campus, bringing in an average of 400,000 dollars per faculty member each year (Clarke and Baumgart 2004). Achieving a high level of funding for pursuing credible research, a quest that had been initiated by Simonett, demonstrated not only the program's success, but also the effectiveness of the transformational leadership that had characterized the founding chair.

In essence, though several noteworthy leaders chaired geography programs in the twentieth century, since the establishment of the graduate program at Berkeley in the 1920s, no geography department chair exhibited the level of transformational leadership that characterized Carl Sauer—until David Simonett took the helm at Santa Barbara. As Sauer sparked a new paradigm in American geographic thought, so did Simonett. His impact has been pervasive. Like the wise sage at Berkeley, Simonett's persistence, inspiration, and vision fueled remarkable changes in geographical thinking and scholarship worldwide. In fact, as Waldo Tobler indicated to me in conversation, he articulated a vision and identified an objective with the sense that "this is the way we had to do it to get it done." His colleagues and subordinates bought into the vision as well as the overall strategy for its implementation, and made the commitment to work harder to achieve the program's goals.

Akin to an earlier time in the discipline's history when the study of regions emerged prevalent in many programs, toward the end of the century, technology, largely in the form of GIS, had become an integral component of geography curricula. And it's fair to say that Santa Barbara supplanted most other departments, including Berkeley and UCLA, as a premier graduate program in geography. This outcome can be considered an unquestionable result of the transformational leadership exhibited by David Simonett as the founding chair of the Santa Barbara department.

NOTES

1. The title for this chapter makes special acknowledgment of David Simonett's gift as a wordsmith. It is taken from a paper I presented at the annual meeting of the California Geographical Society in 2013, which drew the attention of the son of Simonett who happened to notice it while surfing the Internet, and he commented on his father in an August 15, 2013, e-mail: "Thank you for acknowledging his hard work. We chuckled at your creative use of words in the title. He was a wizard with words and I enjoy/suffer from luciferous logolepsy (google that one) because of his 'sometimes off' sense of humour. I can still hear him say 'bloody fahtid dingo kidneys'. Try googling "Simonettisms" to hear the mad Aussie's words. I suspect you might've already. . ."

2. It would have been difficult to find a more distinguished geographer that was professionally active in the latter part of the twentieth century who was advocating for the discipline. De Blij was awarded the doctorate at Northwestern in 1959, completing a solid dissertation on Swaziland's physical geography, and subsequently served on the faculty of Michigan State University and the University of Miami. De Blij's record of research and publication through the years was admirable, as was his leadership in the discipline as an editor devoted to geographic education, geographic research, and department chair. Yet, the kind of geography De Blij practiced was not the kind that the hiring committee desired.

3. While delivering a paper on leadership in American geography at the annual meeting of the National Council for Geographic Education in 2006, I related this story to a full house of academic geographers using my best Aussie brogue where apropos, when in the back of the room, a heavy-set gentleman shouted, "That's *exactly* how it happened!" It was Reg Golledge who made the remark. I had never met him before, and after my presentation on the emergence and decline of American academic departments of geography, he was among the several individuals that approached me, expressing appreciation for my work. As a former President of the Association of American Geographers, he stated that my work needed to be published; it was imperative for the community of academic geographers to read my assessment of leadership

throughout the discipline's history. Susan Hardwick, a former President of the National Council for Geographic Education agreed, and Mary Lynne Byrd, the Executive Director of the American Geographical Society also concurred, and they also were among those that asked when the work would be published. It was at that point when my intention shifted from completing a short article to writing a multi-chapter book.

4. Simonett's widow related to me how he was indeed concerned about his students' well-being. During the Tiananmen Square protests of 1989, his three Chinese students were worried about their families. David Simonett took 30,000 dollars of his own funds to see that their families could leave China. As far as his wife knew, the students were not made aware that he used his personal funds to make this happen.

5. Evidence that GISc technology has been embraced by practitioners of the discipline of all species is pervasive; perhaps most significant is the 2012 Smithsonian American Ingenuity Award to Anne Kelly Knowles, Professor of Geography at Middlebury College, who has employed GIS in her studies in historical geography (Horwitz 2012).

6. Waldo Tobler once related to me accounts of Simonett's persistent character, and he commented that he had no regard for time of day. Simonett would telephone his future colleagues in the middle of the night in efforts to convince them to join him in the fledgling department in Santa Barbara. And join him they did.

7. A 1969 national ranking placed the geography program at Michigan second and Ohio State seventh.

8. Tobler indicated to me that Simonett chose colleagues that were inclined to value the technological applications of geography; yet, given their many accomplishments, both individually and as a collective unit, Simonett must have recognized their potential to develop a rapport with one another and do great things.

TWENTY

The Transformational Leadership Imperative

A number of leaders have been discussed in this work as having shaped the course of the history of geography, and their leadership behaviors have been assessed in reference to five components of transformational leadership—*inspiration, integrity, selflessness, scholarship,* and *proaction*. Table 20.1 shows whether or not each leader is characterized by one or more of these transformational leadership behaviors. Those twenty-two that have been classified as transformational leaders all showed evidence of high levels of *inspiration, integrity, selflessness, scholarship,* and *proaction*. In addition to the twenty-two transformational leaders, nine transactional leaders, five laissez faire leaders, and six *enigmatic leaders* have been determined.[1]

The *enigmatic leader*, a term coined here, refers to the leader that is not characterized by a preponderance of leadership behaviors of any single type. The *enigmatic* leader is not known for his/her leadership attributes; he/she typifies those cast in quasi-administrative leadership roles in higher education today, such as the rotating department chair. It should be noted that each of the transformational leaders showed evidence of carrying the transformational leadership characteristics; however, in all cases each university administration provided support, whether financial or otherwise. Moreover, in each of the cases nearly all, if not all, faculty in the program cooperated with the chair to achieve a common vision. If these conditions were present during the administration of an enigmatic leader, perhaps they would be regarded as transformational leaders as well. Having said that, the leadership styles of Finch (Wisconsin), Trewartha (Wisconsin), Smith (Ohio State), Jones (Northwestern), McClellan (UCLA), and Zierer (UCLA) exemplify that of the enigmatic leader. Though each exhibited certain transformational leadership behaviors, to classify each as a transformational leader is unwarranted, for at least one of the essential ingredients necessary for transformational leadership to occur is lacking from each of their profiles. It is argued that given the current trends in American academe for departmental chairmanships to rotate among faculty members in each department, enigmatic leadership is the leadership behavior typically characteristic of departmental chairpersons today.

The transactional leaders identified include: Clark (Wisconsin), Robinson (Wisconsin), Hartshorne (Wisconsin), Broek (Minnesota), James (Syracuse), Bowman (Johns Hopkins), Carter (Johns Hopkins), Atwood (Clark), and McCarty (Iowa). It must be noted that the moral imperative requisite for transformational leadership is at least to some degree absent in transactional leadership. All transactional leaders listed in Table 20.1 showed evidence of strong scholarship; but in all of these cases, some level of selflessness has been called into question, and the matter of contingent rewards has been considered important. Still, prior to the 1970s, the practice of transactional leadership likely was the norm in American academe. Moreover, although some evidence points to Carter, Bowman, and Atwood as being characterized by some level of malevolence, James, Clark, Robinson, and Hartshorne simply appeared to practice the leadership model that was typical of their time.[2]

Whitbeck (Wisconsin), McMurry (Michigan), Davis (Minnesota), Whittlesey (Harvard), and Martin (Washington) show evidence of having demonstrated laissez-faire leadership behaviors. Here, inspiration and proaction attributes were noticeably absent among each of these faculty cast in the role of department chair.[3] In reference to the list of transformational leaders presented in Table 20.2, of the sixteen programs, all but four (Harvard, Iowa, Wisconsin, and Michigan) had experienced transformational leadership at least once during the department's history. At Harvard and Michigan, mediocre leadership facilitated the termination of programs, Iowa has never been in the "Top 10," and Wisconsin is an anomaly, whose success can be explained by large program size and the likely presence of informal transformational leaders, tirelessly working behind the scenes to facilitate program success. In the case of Chicago, a long line of transformational leaders served as chairs from the founding of the geography graduate program in 1903 until 1960, when Gilbert White left the chairmanship. Berkeley also was characterized by transformational leadership for quite an extended period, first under Sauer for more than thirty years, and then another decade under his former students and later colleagues, Leighly and Parsons. Of the twenty-two transformational leaders, as Table 20.3 indicates, the majority claimed academic ancestral origins to Rollin Salisbury; four of them were students of Sauer.

The transformational leaders at Minnesota took degrees at both Wisconsin and Minnesota, which were not known as environments charac-

Table 20.1. Transformational Leadership Attributes among Leaders in Academic Geography

		Inspiration	Integrity	Selflessness	Scholarship	Proaction
Chicago	*TF* Salisbury	Y	Y	Y	Y	Y
	TF Barrows	Y	Y	Y	Y	Y
	TF Colby	Y	Y	Y	Y	Y
	TF Platt	Y	Y	Y	Y	Y
	TF White	Y	Y	Y	Y	Y
Wisconsin	*LF* Whitbeck	N	Y	U	N	U
	E Finch	N	Y	N	N	U
	E Trewartha	M	Y	U	Y	U
	TA Clark	Y	M	N	Y	U
	TA Robinson	Y	M	N	Y	U
	TA Hartshorne	M	M	N	Y	U
Michigan	*LF* McMurry	N	U	U	M	U
Minnesota	*LF* D.H. Davis	N	N	N	N	U
	TA Broek	N	U	N	Y	U
	TF Borchert	Y	Y	Y	Y	Y
	TF Lukermann	Y	Y	Y	Y	Y
Ohio State	*E* G.H. Smith	M	Y	U	M	M
	TF Taaffe	Y	Y	Y	Y	Y
Penn State	*TF* E.W. Miller	Y	Y	Y	Y	Y
Syracuse	*TF* Cressey	Y	Y	Y	Y	Y
	TA James	M	U	M	Y	Y
Clark	*TA* Atwood	M	N	N	Y	Y

	TF Cohen	Y	Y	Y	Y	Y
Johns Hopkins	*TA* Bowman	M	M	M	Y	Y
	TF Lattimore	Y	Y	Y	Y	Y
	TA Carter	M	N	N	Y	Y
	TF Wolman	Y	Y	Y	Y	Y
Harvard	*LF* Whittlesey	M	U	Y	Y	N
Northwestern	*TF* G.D. Hudson	Y	Y	Y	Y	Y
	E Jones	M	U	U	Y	Y
Washington	*LF* Martin	N	U	U	N	N
	TF G.D. Hudson	Y	Y	Y	Y	Y
Iowa	*TA* McCarty	M	N	U	Y	M
UCLA	*E* McClellan	Y	U	Y	N	Y
	TF McBride	Y	Y	Y	Y	Y
	E Zierer	U	U	U	N	U
	TF Spencer	Y	Y	Y	Y	Y
	TF Bruman	Y	Y	Y	Y	Y
Berkeley	*TF* Sauer	Y	Y	Y	Y	Y
	TF Leighly	Y	Y	Y	Y	Y
	TF Parsons	Y	Y	Y	Y	Y
Santa Barbara	*TF* Simonett	Y	Y	Y	Y	Y
	TF Golledge	Y	Y	Y	Y	Y

Note: Transformational Leadership Attributes: *Inspiration, Integrity, Selflessness, Scholarship, Proaction.* See Appendix B for documentation.

Key: Y = Yes (this leader is characterized by this attribute); N = No (this leader is NOT characterized by this leadership attribute; M = Moderate (this leader is moderately influenced by this leadership attribute); U =

Unknown (it is unknown whether or not this leader is characterized by this leadership attribute; *E* = Enigmatic Leader; *TF* = Transformational Leader; *TA* = Transactional Leader; *LF* = Laissez-faire Leader

terized by transformational leadership at the time. One might speculate that their leadership attributes were fostered by service in the armed forces; however, there appears to be no correlation between military service and transformational leadership behaviors. Though Borchert was marked by this kind of experience, several other leaders with similar experiences showed transactional leadership behaviors. For example the trio of autocratic oligarchs at Wisconsin all served in the OSS, and Whittlesey, the laissez faire leader at Harvard, served as both an enlisted man and an officer during the First World War. On the other hand, two of the most influential transformational leaders, Sauer and Simonett, had no military or government intelligence experience at all. Sauer's only non-academic employment was in journalism and cartography. Simonett was a bank clerk in Australia during World War II (Kenzer 1985; Leighly 1976; Morain 2006).

Two of the transformational leaders were targets of McCarthyism: Lattimore, for sure, and Cressey, slightly less so. Both had considerable experience in China and commented without then-political correctness on actual conditions and circumstances, rather than voicing party-line views. Cressey's overseas experience was related to his role as a missionary educator, as was McBride's in Latin America; thus, two transformational leaders claimed extensive foreign area field experiences sparked by Protestant evangelization efforts. Yet in both cases it appears that rather than acting as proselytizers of a belief system, Cressey and McBride concentrated more on learning about the people and the land through interaction with them, and they conducted sound scholarship that lent great insight to many. Though Cressey was able to oversee the development of a doctoral program after the Second World War, McBride's call to service for the government during the war precluded that opportunity, for the Ph.D. program at UCLA was not initiated until his retirement; however, it appears his influence in developing the transformational leadership behaviors of his undergraduate students, Spencer and Bruman, was nearly equivalent to the influence Sauer had in developing them as leaders in the field.

Three sons of professors took on transformational leadership roles: Sauer, Lattimore, and Wolman. The majority of transformational leaders pursued undergraduate studies at small institutions, such as normal schools, former normal schools, or small independent colleges. All engaged in some sort of field studies at one point in their career, and most developed some international experience. Not bound to traditional classrooms, this element of their education and professional growth played an

Table 20.2. Transformational Leaders in American Geography

Chicago	Salisbury
	Barrows
	Colby
	Platt
	White
Minnesota	Borchert
	Lukermann
Ohio State	Taaffe
Penn State	E.W. Miller
Syracuse	Cressey
Clark	Cohen
Johns Hopkins	Lattimore
	Wolman
Northwestern and Washington	G.D. Hudson
UCLA	McBride
	Spencer
	Bruman
Berkeley	Sauer
	Leighly
	Parsons
Santa Barbara	Simonett
	Golledge

important role in broadening their horizons and shaping their mental models of scholarship and leadership.[4]

An attribute often not explored closely enough in examining leaders relates to their upbringing, and here this must be addressed. Whereas leaders like Sauer, Cressey, McBride, and White were raised to embrace a high level of social consciousness, this should be examined further as a possible key indicator for transformational leadership.[5]

Lastly, of these twenty-two stellar leaders, eleven were cast in the role of influencing the discipline's development at the departmental level with cooperation and support from the academic administrative leadership in founding new programs, resuscitating programs, or creating marked changes in already existing programs through selective hiring of new faculty: Sauer (Berkeley), Simonett (Santa Barbara), Salisbury (Chicago), Hudson (Northwestern and Washington), Taaffe (Ohio State), Cressey (Syracuse), Miller (Penn State), McBride (UCLA), Borchert (Minneso-

Table 20.3. Transformational Leaders in American Geography Claiming Academic Lineage from Salisbury

Barrows
Colby
Platt
White
G.D. Hudson
Cressey
Sauer
Spencer
Bruman
Leighly
Parsons

ta), Wolman (Johns Hopkins), and Cohen (Clark). This matter of administrative support in facilitating departmental program development appears to be an imperative in the implementation of effective transformational leadership; but it takes an exemplary individual marked by *inspiration, integrity, selflessness, scholarship,* and *proaction,* to articulate a vision that generates commitment from the administration and faculty colleagues to seek to achieve excellence.

In essence, it is leadership behavior that has been influential in the development of programs that markedly shaped the study of academic geography in America.[6] The conditions for each program's development have differed; however, it is clear that under laissez faire leadership, the threat of demise is most apparent, and not only does greater likelihood of success occur when transformational leaders are in positions of authority, their success seems to continue far beyond their terms as formal leaders. Moreover, though transactional leaders have played important roles in some programs, their influences in the development of successful programs have been inconsistent and appear to be of short duration. No doubt, external factors also create marked impacts on the evolution or devolution of academic programs, and the degree to which support is drawn from administrative deans, provosts, and presidents is significant, for without this kind of backing wheels are often left spinning. *Still, the evidence is quite clear that if academic departments and their host institutions desire to develop programs as centers of excellence, transformational leadership is an imperative.*

NOTES

1. With the exception of Isaiah Bowman, because he had a special relationship with the administration of geographical study at Hopkins, only leaders that served in official capacities as program directors or chairs at some point in the geography program's history have been examined in this study.

2. Not enough is known about McCarty and Broek's leadership behaviors to determine whether or not either sought to be intentionally injurious to a colleague or a subordinate. The same cannot be said of Carter, Bowman, and Atwood.

3. It must be noted here that although Whittlesey's leadership in the department was laissez faire, he effectively mentored students. Cohen remarked that Whittlesey "took me under his wing. He was a thoughtful, courteous person—and insightful. I learned a great deal from him." It cannot be emphasized here enough that Cohen became one of the most influential transformational leaders in the history of American geography.

4. An exception here is Saul Cohen, who attended Harvard as an undergraduate. Incidentally, he related to me that he learned much about leadership as a faculty member at the Naval War College.

5. It might also be useful to examine a marked absence of social consciousness in relation to the development of transactional leadership behaviors. It appears likely that the leader raised with a sense of entitlement would have difficulty adopting transformational leadership behaviors. Incidentally, Simonett, too, was raised to embrace a high level of social consciousness. His brother was mentally handicapped, and certainly this drove his mother to work tirelessly on behalf of the mentally disabled. In fact, she was decorated by the Queen for her significant contributions. The point is that it was in this sort of environment—one that evoked an obligation to care for the disenfranchised—that characterized David Simonett's upbringing as well.

6. Academic disciplines other than geography can learn much by adopting a similar historical approach to examining how leading programs have been shaped by their formal leaders.

TWENTY-ONE

Epilogue

In retrospect, it is evident that the historical narrative continues to play an important role in explaining the conditions, nuances, and circumstances that facilitated the emergence of past transformational leaders in the geographic discipline, as well as uncovering the blunders of those marked by laissez faire behaviors. As the leaders of the discipline's most influential programs have been identified and evaluated in a manner that heretofore has not been engaged in by historians of the geography, the matter of employing leadership as an important context for developing an understanding for the history of science has been shown to be an essentiality. Further, as all scientific disciplines undergo evolution, acknowledging that understanding the paths followed in the past is imperative for recognizing the appropriate paths to take in the future is imperative. Moreover, as leaders often dictate the course of a discipline's destiny, it is the paths they have chosen, often largely as a consequence of their inherent leadership behaviors, which determine the options from which their followers and subsequent leaders may choose.

In essence, throughout the history of geography in the United States, leadership has played a significant role in both the success and failure of programs and departments tasked with advancing the discipline. Leaders characterized by a heavy dose of transformational leadership attributes, such as Salisbury, Sauer, and Simonett, sparked success with substantial support drawn from the administrative leadership in their respective institutions. The same situation occurred with G.D. Hudson at the helm of two separate departments, as well as Taaffe, E.W. Miller, and Cohen. With support coming from administrative deans, willing to engage in collaborative efforts to fuel the success, geography indeed triumphed, as the department chair was able to hire his own (faculty) subordinate colleagues. Careful selection of departmental members enabled these transformational leaders, cast in the role of departmental chairmen, to determine likely candidates willing to invest in a common vision for the future, which although it underwent evolution over time, he defined. Consequently, flagship programs emerged on the campuses of Chicago, Berkeley, Santa Barbara, Northwestern, Washington, Ohio State, Clark, and Penn State.

That being said, the case of Wisconsin is unusual because there is no evidence that transformational leadership behaviors dominated among its chairs over its long legacy as a member of the bellwether programs in geography; nor were conditions necessarily optimal for transformational leadership strategies to be implemented by the department's administrative leaders. Here Trewartha's introduction of rotating departmental chairmanships, perhaps inspired by a combination of his sense of democracy as well as distaste for administrative duties, fostered enigmatic leadership, and subsequent chairs vacillated back and forth between enigmatic and transactional leadership attributes. Yet on the campus in Madison, rising stars like Butzer, frustrated by the politics that marked those characterized by transactional leadership behaviors, departed for elsewhere. Others that remained, such as Denevan, appeared to have engaged in transformational leadership, largely as informal leaders "behind the scenes." These informal leaders, characterized by transformational leadership behaviors, mentored students and junior faculty alike with high levels of integrity and selflessness, while also engaging in credible scholarship. They served with effectiveness in maintaining the department's flagship status; yet, as with elsewhere, the program's large size also likely enabled it to mask haughty prima donnas disinclined toward contributing more than their scholarship to the synergy that maintained its elevated ranking. This was a burden left to other colleagues that recognized the necessity of sharing duties integral to the sound functioning of academic departments; but it must be noted that even in such a large department, with the retirement of the oligarchs, the presence of collegiality became more of a norm than its absence.

Less than a few hundred miles northwest of the campus in Madison, the University of Minnesota's department was presided over by a tyrant during its early years, whose insecurities led to behavior that motivated the departure of leading scholars. Following D.H. Davis's tenure as chair, came the brief, autocratic reign of Jan Broek. Subsequently, transformational leaders, characterized by self-confidence and independence in spirit, appeared as departmental chairs inspiring colleagues and students alike. Borchert and Lukermann were notable in this regard, and through their efforts as departmental chairmen and mentors, the geography department in the Twin Cities became known as one of the best.

Also in the Midwest, Michigan commanded a significant presence in the geographical community; however, this is more likely a result of things external to the program's leadership, for throughout its history, with the exception of Sauer, its chairs appear to have been relatively inconsequential to the successes it enjoyed. Prior to World War II, the discipline was housed in only one of the few independent departments of geography in American academe, and fewer still were the number of graduate programs. Thus, the simple fact that it awarded doctorates during the first half of the twentieth century lent the program credibility. Moreover, by being able to claim an affiliation with one of the country's premier public universities, its status was enhanced as well, which was boosted by the presence of world-class scholars in the 1960s and 1970s. Yet, an absence of leadership accompanied the loss of senior faculty, such as Tobler, who left for a sunnier climate, and the department subsequently folded. Here, it can be argued, the department's members and its chair(s) showed no evidence of engagement with the larger university community; further, the department had not created the liaisons necessary throughout the institution to support the notion that geography on the campus in Ann Arbor was a subject suitable for study in an American university.[1] Laissez faire leadership—and followership—appeared to characterize the departmental culture from time to time after Sauer's departure in the early 1920s. For an academic discipline as seemingly as esoteric as geography, it is then little wonder that in a time of questionable financial solvency at the state-level, the department met its demise in the early 1980s.

Perhaps at an institution as prominent as the University of Michigan, the demise of geography at Harvard played a critical role, for only Berkeley surpassed the Ann Arbor university in national rankings of public universities, and on occasion it was tied by UCLA; however, if Harvard did not have geography, was it a viable university subject? Harvard had lost its geography program as a consequence of a few things that when taken together, facilitated its termination. Although its famous alumnus and (then) Hopkins President, Isaiah Bowman, could have done more to save the department, Harvard's demise largely came as a result of Whittlesey's laissez faire leadership.[2] There is no real evidence to indicate that in Ann Arbor a similar situation did not occur. After its program folded, geography was terminated at Chicago and Northwestern—with no evidence to indicate that the state of affairs concerning leadership was substantially different. Only Johns Hopkins remained among elite private universities that supported a geography graduate program, and then it was compelled to withstand a legacy of transactional leadership and treachery. It never joined the ranks of the "Top 10," and the presence of geography within the Department of Geography and Environmental Engineering became increasingly less significant as the century neared its close.[3]

Syracuse and Clark, second-choice schools to the Ivies,[4] remained as the only significant graduate programs in geography housed in private institutions. Clark engaged in the mass production of geography doctorates during Atwood's presidency of the institution and administration of the geography program. Arguably, in the long run, his transactional leadership did the discipline little good.[5] Clark did not appear in the "Top 10" until the 1980s, and its success can largely be attributed to the transformational leadership of Saul Cohen. Syracuse had gained elevated status largely as a consequence of Cressey's transformational leadership early on, as well as the positive image James built for the department, by assembling a group of stellar scholars that played informal leadership roles. For example, the presence of Donald Meinig, certainly one of the leading geographers in the English-speaking world, stood the department in good stead. This was particularly relevant at a time in the discipline's history when the historical geography embraced by Meinig was considered *passé* by many during an era when numeracy was the fashion.

Quantitative methods gained prestige at its places of origin in Seattle and Iowa City throughout the four decades following World War II. Interestingly, the program at the University of Iowa, made famous by Schaefer and his article that rebutted Hartshorne's treatise on the nature of geography, like Hopkins, never entered the "Top 10." Iowa did, however, gain significance as an influence in the development of the discipline simply because it sparked debate and fueled the quantitative revolution that fostered a different way of thinking among mostly younger members of the discipline who were not, by and large, accepted by geography's establishment. McCarty, the program chair, was embarrassed by the debate, and he sought to distance himself and the program from Schaefer's philosophy, which met with many favorable reviews among those disinclined to accept the traditional practice of the field. Perhaps had he not practiced transactional leadership, the department would have grown to include larger numbers of the discipline's practitioners, and then it might have gained elevated status. Unfortunately, that is not what happened. One of the most significant programs in the discipline's history was relegated to the second tier, where it remains.

Lastly, UCLA's history is marked by a mixture of transformational leadership and that which has been enigmatic. Early on, the transformational leadership practiced by McBride created conditions for a solid graduate program. His students, Bruman and Spencer, took doctorates at Berkeley and ultimately returned to Los Angeles, where they practiced transformational leadership behaviors that emulated both McBride and Sauer. The graduate program flourished early on, a marked consequence of stellar leaders, and in five national rankings between 1957 and 1995, it never appeared outside the "Top 10." It grew to great size, and by the end of the twentieth century characterized the diversity practiced by geographers in several different fields. Although the enigmatic leadership that appears to be the norm in academe today characterizes its recent chairmanships, it was its history and its large number of faculty that have been devoted to substantive scholarship that have contributed much to its success. Moreover, it is a department, which unlike several programs, has shown a history of consistently not being reluctant to award doctorates to many of its deserving students.[6]

In essence, in reference to the programs that influenced its history, the practice of transformational leadership in American geography has effectively shaped the discipline's development by fostering success in its most stellar programs. In contrast, laissez faire leadership has been a recipe for disaster. Transactional leadership has met with mixed results; departments that have experienced success where chairmen have been characterized by transactional leadership attributes, informal transformational leaders played significant roles. Lastly, the enigmatic leadership that more frequently characterizes departmental chairs in academe today generally is successful in maintaining the status quo. The results are obvious. If the discipline seeks to move beyond the existing state of affairs in its academic departments, transformational leadership becomes increasingly an imperative in American geography.

DIRECTIONS FOR FUTURE RESEARCH

It is the historical perspective that establishes the foundation for future research, and here the options are many, for this study has raised several questions that are worthy of exploration. For example, although this study has shown that a number of transformational leaders begot transformational leaders,[7] transformational leaders also often failed to produce transformational leaders. Though Sauer diffused transformational leadership characteristics to Spencer and Bruman, as well as Parsons and Leighly, his student, George Carter, appeared to be marked by transactional leadership attributes as well as questionable ethics. The question then arises as to whether or not the graduate student's mentor plays an important role in shaping his or her prospective leadership behaviors. Further research here is needed to evaluate the influence of the academic mentor on affecting not only the student's scholarship, but also his/her ethical character and prospective leadership behavior.

One of the problems in American graduate education is that students are trained to become scholars, not leaders—and not necessarily followers; however, the higher education institutions that act as repositories for Ph.D.s are marked by various institutional units, such as departments, programs, schools, and colleges, which require stellar and well-prepared leaders to administer them, but often are characterized by former professors of English, biology, or psychology cast in leadership roles and unprepared for leadership positions. They often have no training or no experience in higher education leadership; yet they are tasked with supervising talented professionals and directing them to become stakeholders in the institution's mission, frequently fluid, as defined by transient university presidents. Consequently, poor leadership in higher education is not a seldom occurrence, and it far too frequently results in an alienated group of followers. Hence, the enigmatic leader as chairperson appears to be established as the rule—and an acceptable norm at that. In an academic environment marked by a high level of autonomy practiced by its faculty, enigmatic leadership might be the easiest, but not the best, leadership behavior to practice. Here it must be reemphasized that although life experiences might influence leadership behavior, transformational leadership attributes can be learned (Bass 1990; Avolio 1994). Among other things, given that transformational leadership behaviors are shown to spark organizational commitment, transformational leadership becomes an imperative for the success of academic departments and programs (Avolio et al. 2004).

This being said, it stands to reason that if one is to operate under the premise that many doctoral students may be cast into higher education leadership roles, some compulsory leadership training in Ph.D. programs is necessary. Thus, research devoted to examining the leadership training of higher education leaders is important so that it can be determined what the proper forum might be for facilitating this type of leadership education.

Moreover, the matter of ethics in both administering subordinates and colleagues as well as engaging in scholarship must also be addressed. Graduate students do not receive an adequate base of knowledge and training in higher education ethics *vis-à-vis* not only research and publication, but also with regard to administrative leadership. In a society marked more by contingent rewards rather than the less commonplace efforts devoted to the larger purpose of contributing to a greater social good, the importance of selflessness and integrity must be addressed in graduate education. Thus, some education at the graduate level, which incorporates the necessity of adherence to a particular code of ethics as well as the importance of higher education faculty to take on roles as servants of society, is important as well. It is argued that coursework demonstrating proficiency in higher education leadership for doctoral students in the liberal arts and sciences is not an unreasonable requirement, and this should include modules in leadership, legal issues, and ethics. Some research in developing and implementing this kind of curriculum is much needed.

More research also is needed on the experiential phenomena outside academe, which have played influential roles in the lives of leaders *vis-à-vis* their selection of leadership behaviors. For example, it might be reasonably speculated that hierarchal organizations, such as the armed forces, prepare individuals for leadership positions in higher education. Yet, this study has shown that supervisory and command experiences in the armed forces and government intelligence do not necessarily correlate with sound leadership in higher education.[8] In fact, the transformational leaders at the top of the list, Sauer and Simonett, had no military or government leadership experience, whereas laissez faire leadership was experienced by Whittlesey, who served in both the enlisted and officer ranks, and transactional leadership was exhibited by the Wisconsin OSS veterans: Clark, Robinson, and Hartshorne. Thus, it would be prudent to study transformational leaders more deeply with regard to the forces that have affected their choice of leadership behaviors, for if certain influences (e.g., overseas missionary work) tend to foster these preferred leadership attributes, then screening and recruitment of prospective faculty leaders should take these experiences into account.[9]

Also direly needed is study on informal leadership in academe. Academic departments sometimes lend themselves to laissez-faire leadership because the chairmanship is often merely an administrative position largely devoted to facilitating dialogue with other elements of the university and engaging in the bureaucratic functioning of the academic unit. As matters associated with accommodating deans, provosts, and presidents' demands for data and memoranda consume much effort that takes away from scholarship and teaching, the chairmanship is frequently avoided by stellar informal leaders who actually serve as the true leaders in the academic department. They may be leaders in the discipline through scholarship and mentoring, as well as leaders behind the scenes providing nearly endless bouts of counsel to department chairs, often crafting major decisions affecting the program's future; however, though their impact is significant, and their leadership may be recognized as convincing, it holds no official status. Especially in universities, their roles are compelling, and here leadership studies must take hold. For example, the influences in the discipline of figures like Donald Meinig, David Harvey, Peter Gould, and Robert Kates have been noteworthy, their leadership roles have largely been in informal capacities. They are not alone.

This study has not included assessments of programs in the second tier, or those that do not offer doctorates; thus, an examination of departmental leadership in programs lacking prestige might be in order, because they constitute the norm. It is arguable that with the passing of greats and the emergence of GIS, parity is more closely realized; the recent NRC rankings show the University of Idaho as potentially in the "Top 10," along with a host of some sixteen other doctoral programs (Doctoral 2010). Legacy, though still important, has appeared to have lost ground in assessing the perception of a program's quality. Despite the lack of value of the recent rankings, one thing is clear; the quality of geography programs appears to be becoming more similar than unlike, and rankings seem less important. Perhaps the most useful ranking, which has yet to be employed, is one that solely relies on calculating the number of each department's Ph.D. graduates employed as faculty in American geography doctoral programs, for it is the doctoral program that is assigned the duty of providing the foundation and training of academic geographers, and endorsing their membership to the academy.

It is in geography programs at all levels that additional research might be taken to assess current trends and perceptions in transformational leadership. The multifactor leadership questionnaire (MLQ) can be applied to assess the perceptions of program chairs' leadership behaviors of themselves, as well as their followers perceptions of them. This procedure is widely accepted and has been used extensively in a variety of settings to develop an understanding of the effects of leadership, especially with regard to measuring transformational leadership, which is postulated as being the most effective leadership behavior (Bass and Avolio 2000). Ample evidence indicates that transformational leadership is preferred as a superior leadership behavior; yet transformational behaviors are lacking among department chairs (e.g., Arcuri 2002; Gitttens 1996). Conducting these kinds of studies not only offers much needed contributions in scholarship, they also provides much-needed input to department chairs with regard to their leadership behaviors and spark changes that may enable them to become more effective leaders.

REFLECTIONS AND CONCLUSIONS

It was not too long ago at a conference of academic geographers, I attended the department chairs' breakfast. At this gathering of geography program heads from throughout the region, much was discussed, but it had very little to do with geography. Instead, discourse became sharply focused on dismay and diatribes targeting adversarial relationships that typified the fractures between geography departments and the administrative leadership of the chairs' respective institutions. Certainly, this contentiousness between higher education leaders and their subordinate faculty is typical, and it frequently appears to be marked more by an armistice with an undercurrent of hostility, rather than true peace, reconciliation, and collaboration in decision-making for optimal achievement of the institution's goals. Yet, both administrative leaders and faculty, including departmental chairs, are tasked with taking strides to denude the hostility and working toward a common effort to achieve their respective institutions' overall mission.

The department chair must foster these efforts because support from the administrative leadership is imperative for success. This is clearly apparent in the history of American geography; Sauer, Simonett, Salisbury, G.D. Hudson, E.W. Miller, and Taaffe, for example, all received strong support from their institutions' leadership to advance geography graduate programs upon their hiring. Granted, in certain cases, an institution's leadership is a lost cause and a major accomplishment might be in the maintenance of the status quo until a new generation of administrative leaders appears. This was the situation that Simonett faced at the University of Sydney in Australia when he found the situation intolerable and decided to leave for a position in private industry. Although it is the department chair's responsibility to convince the administrative leadership that the program's vision is integral to the institution's mission, there are things that are beyond the chair's control, and the whims of deans, provosts, and presidents are included in that category. Simonett's case is a most telling example of the need for administrative support, for his departure from Sydney was their loss and Santa Barbara's gain. With the opportunity to pursue his vision for a new kind of geography program in California, which was compatible with the institution's strategic plan, coupled with the freedom to hire people he personally selected, Simonett's transformational leadership was facilitated by the university administration.

This matter of the department chair being given the opportunity to screen and recruit his/her own faculty colleagues appears to be an especially important criterion for the transformational leader to achieve the utmost success. For example, Sauer, and Simonett, were given *carte blanche* in hiring prospective faculty. Once their programs were built, chairs that had apprenticed under these transformational leaders took charge of their respective programs (e.g., Leighly and Parsons at Berkeley and Golledge at Santa Barbara), and acted as transformational leaders themselves; however, more recently, transformational leadership appears to have vacated the premises from which it once thrived.

Chicago's program closure in 1987, which marked the end of a legacy, is a case in point. A series of transformational leaders beginning with Salisbury and ending with White brought prosperity to the program. During the 1960s, 1970s, and 1980s, the department, still marked by stellar academic geographers, maintained a high ranking (in 1969, the year before White departed for Colorado, the department placed number one nationally); however, its small size, internal discord, and ineffectiveness in justifying its existence to the administration brought it to demise.

Transformational leadership had fueled its existence; but unlike other geography graduate programs in the "Top 10," Chicago was not large enough to argue for its viability when the transformational leadership that characterized a long line of chairs no longer prevailed, and this was especially problematic when a malcontent's voice became increasingly vocal. This was not the only case in our discipline's history where "one bad apple spoiled the whole bunch."

On the other hand, the graduate program at Wisconsin, for example, not only maintained its viability because of the scholarship produced by its noted membership, its large size, in terms of faculty (and students), lent credibility and strength. Thus, more recently, in the absence of transformational leadership, program size appears to play an important role in maintaining each individual program's existence—and status.[10]

Program size also increased the likelihood of informal transformational leaders serving in departments, contributing to scholarship, and mentoring students, thus enhancing the graduate program's prestige. William Denevan at Wisconsin, a student of Parsons, and Jonathan Sauer at UCLA (the son of Carl), are likely examples, and their efforts in conducting cross-disciplinary research appear to foster a higher regard for the discipline, as well as the department, in the greater community of scholars.

As membership to this community is generally designated by the award of the doctorate, it is worth reexamining the notion of increased parity between and among departments. Doctoral programs that emerged at the turn of the most recent century, such as San Diego State University and Texas State University, are finding placement of their alumni in academe increasingly becoming more likely as they develop reputations for sound teaching and meeting the needs of the academic marketplace. As even newer programs have appeared at Nevada-Reno, Southern Mississippi, and North Carolina-Charlotte, only time will tell as to whether or not the academy will consider them to be credible additions. A scanning of the Association of American Geographers (2010) directory shows a relatively strong representation of Ph.D. alumni from the leading programs on the 1995 "Top 10" list in geography doctoral programs today. This indicates a reality that has far-reaching ramifications, for there appears to be a general preference for hiring new assistant professors from what are perceived to be more prestigious programs and a lack of desire to recruit faculty from more "pedestrian" programs. Showing a frequent absence from the rosters of faculty in geography doctoral programs are Ph.D. graduates from third tier programs that have been in existence for four decades or more. A number of these departments have failed to move beyond their status quo of mediocrity, and this constitutes a failure in leadership.[11] Here transformational leaders are much needed.

Moreover, the application of geographical learning to prospective employment also is a key ingredient to successful programs.[12] Jobs for graduates—whether in academe, government, or private industry—are an imperative. Successful programs in their infancy have had transformational leaders assist students in gaining employment, and by doing so, have developed strong levels of commitment from their followers toward achieving well-articulated visions. Inspiring followers to pursue a vision is essential, and that component of leadership has clearly been absent in lackluster programs.

Lamented far too frequently is the matter of the Harvard demise, the subsequent absence of the discipline in the Ivy League, and the notion that geography is neither appropriate, nor is it acceptable, in American higher education.[13] This last contention, that the content and methodology of the geographic discipline rendered it a subject not valid for university study and subsequently led to its downfall in American academe, is not necessarily true. Instead, it appears that the discipline met its demise in programs characterized by questionable leadership, and it thrived in departments marked by transformational leaders.[14]

Effective transformational leadership is an imperative in the chairmanship of academic departments, and though this behavior is often innate, it also can be learned with effective results (Bass 1990). Given that, if leadership training devoted to enhancing transformational leadership attributes among geography department chairs is not incorporated by higher education institutions and professional societies, the likely placement of unsuitable faculty in leadership positions becomes increasingly real. In lieu of such training, universities should consider it a priority to employ individuals that characterize these behaviors, and some screening to assess optimal leadership traits may be warranted (Barbuto 2005).

That being said, it must be acknowledged that from time to time one might be tasked with taking on the role of chair in already existing academic programs marked by apathetic faculty engaging in negligent exercise of their autonomy, or even worse, one or more individuals characterized by egotism or narcissism. Here, the herding of cats becomes in the very least, problematic, and the achievement of the department's goals appears only on the horizon of the imagination, like an elusive Shangri-La.

Even worse still are situations in which administrative superiors create stumbling blocks that impede success. It is a paradox of sorts that in academia, academic deans are cast in the role of supervising individuals often more talented, intelligent, and independent than themselves. Thrust into these roles, commonly without the assurance of the job security that comes with tenure, they frequently develop envy or distrust toward their relatively autonomous faculty subordinates, and engage in unnecessary and sometimes harmful actions (or inaction) that impede progress and the achievement of departmental success, which also tend to hinder institutional success. Authoritarian deans, provosts, and presidents are unfortunately not infrequent occurrences in academe, and although their styles of leadership might be preferred by trustees or individuals placed at various levels in an institution's hierarchy of leaders, their leadership behaviors often leave much to be desired. Professors of different ranks may be marked by a host of complexities, for sure, and leading them can be difficult; but, by and large, they are talented professionals, and in a sense, the engine that drives the train of higher education. Engaging in true dialogue with faculty and meeting their needs does more than elevate morale; it serves the mission of the academic institution. Indeed, the onus is on the institution's leaders to embrace transformational leadership attributes and facilitate the development of constructive relationships with faculty subordinates. Yet, too often this is an obligation that is avoided by those steeped in bureaucracy and inclined toward autocracy.

The reality is that faculty colleagues and administrative leaders are too frequently irresponsible and uncooperative, oftentimes selfishly concerned with their own careers more than the university and its mission. In cases such as this, little more can be done than to recognize that the practice of transformational leadership is a goal toward which one must aspire. Here, the *journey* devoted to the craft of leadership becomes just as important, if not more so, than the *conquest*. It is highly likely that in academe, this is the scenario that most often takes place, and enigmatic leadership becomes the normative leadership behavior. Yet, because leadership is essentially a phenomenon revolving around the relationship between leader and followership, the *quest* for transformational leadership, albeit an elusive goal, continues to be an essentiality for both the leader and the led. In academe, the department chair is among those unlikely to realize this, and in a discipline that is as small and seemingly as esoteric as geography, the chair's failure to grasp the nature of leadership threatens the vitality of the discipline.

Thus, in the case of academic geography, it is time for assessment of the history of the discipline within the context of leadership. Here, an understanding of the past lends critical insight to contemporary geography departments, as efforts are made to achieve desired goals, and it is also here that history becomes relevant to all leaders in the discipline.

NOTES

1. Shortly after the closure of Michigan's program, in an address to the East Lakes Division of the Association of American Geographers in Bowling Green, titled: "We ain't got no Image," George Kish, a distinguished professor of Geography at the University of Michigan (1943–1985), lamented about how often he heard that because Harvard had no geography program, there was no reason for the University of Michigan to have one.

2. It must be emphasized here, as indicated in the discussion of Harvard in the text of this work, that among his many accomplishments, Bowman, President of Johns Hopkins University, had been an advisor to two U.S. presidents, he had made the cover of *Time*, and he had been the Director of the American Geographical Society. He also held a number of additional advisory positions on boards of private corporations and government agencies, ranging from AT&T to Annapolis. Bowman, a remarkably effective power-wielder and game theorist—with strong allies in U.S. politics and private industry, ultimately decided against exerting pressure on President Conant at Harvard to save its geography program.

3. As David Harvey reflected on his return to Hopkins before the end of the century, he noted, "By 1993, I was back on the fall line in Baltimore. It was not an easy move. I went from a position of power in a large geography department to being a minor and marginalized figure in a department dominated now by engineers in a very corporatist engineering school that cared only about grants and sponsored research" (Aitken and Valentine 2006, p. 189.)

4. It became commonly known that many undergraduate students desiring to gain admission to Harvard, Yale, or other Ivy League universities, selected Syracuse or Clark as "second choices" if rejection by the Ivies took place, which was often the case.

5. Although the effectiveness of Atwood's leadership is questioned, there is no doubt that the establishment of Clark's geography program was indeed a significant contribution to the field. As the program sought to increase its alumni, it was one of the few that attracted women to the study of geography (Monk 1998). Moreover, the simple fact of it being one of a handful of doctoral programs in the United States created an opportunity for stellar scholars in geography to serve together as colleagues (e.g., Ellen Churchill Semple, Clarence Jones, Samuel Van Valkenburg, Elmer Ekblaw); and distinguished government scientists served on occasion in part-time roles (e.g., O.E. Baker, Curtis Marbut, and Homer Shantz). Having said all this, Atwood failed to attract distinguished luminaries; Charles Colby, Vernor Finch, and Derwent Whittlesey were among those that rejected offers of employment (Koelsch 1988).

6. The matter of doctoral students being treated unfairly in graduate programs has occurred frequently in American academe; many departments have shot themselves in the foot by endorsing unsuitable faculty as graduate advisors and allowing others to engage in misconduct. Far too frequently are cases in which graduate students engage in research and writing and see their work published under the name of a faculty member. If students are fortunate, they may see themselves added as asterisked footnotes, and if they're luckier still, they may be second authors. Yet, they also might see their dissertation research stolen by members of their committees and published without their knowledge in esteemed journals. There are other examples of reprehensible behavior practiced in the world of academe, as this work has shown; however, in comparison with other fields, the geographic discipline is small, and it cannot afford the transgressions too often committed by its representatives in academe.

7. The line from Salisbury to Sauer to Parsons is a case in point.

8. This matter of military veterans *not* exhibiting transformational leadership behaviors in higher education leadership positions is especially disconcerting, and to some degree, surprising and embarrassing. Upon reflection, transformational leadership behaviors were the expected attributes that characterized leaders in my own Marine Corps experience; thus, one must wonder if specific military occupational specialties are more inclined to embrace leadership attributes of a specific category (Transformational, Transactional, Laissez faire, or Enigmatic). Here too, future research should take place, and the results might assist higher education leaders engage in more careful screening of prospective faculty and administrative leaders.

9. There is enough evidence in this study to suggest that past episodes of servant leadership are correlated with effective transformational leadership (e.g., Sauer, McBride, and Cressey). This is an area worth looking into as well, for if individuals with personal histories characterized by a commitment of service to society are seen as likely effective transformational leaders of the future, graduate admissions, assistantship and fellowship awards, and hiring committees might (rightfully) place significant weight on this attribute when making decisions.

10. Although his tenure as department chair was relatively short, David Ward was known as an excellent one; however, the rotating nature of the position compelled one to largely practice enigmatic leadership, though likely with the incorporation of transformational leadership attributes, and at Madison, this was effective. He was asked by friends to consider higher administrative appointments, and in an interview in 2001, he reflected:

"There were three vacancies: Dean of Letters and Science, Dean of the Graduate School, and Provost and Vice Chancellor. My friends wanted me to apply for all these positions. So I did apply for all of them, or people may have nominated me. And there was the usual search, outside candidates, and all of this. I was short-listed for all three, which was a surprise to me because my only experience was as an associate dean in the Graduate School, and there were many deans who were candidates. So I went to see Donna [Shalala], and she said, 'Well, which of these jobs would you like?' I said, 'Dean of Letters and Science.' 'That's not my plan.' So I said, 'Well, you know, that's the one that I'm most qualified for.' She said, 'I think you'd be an ideal Vice Chancellor.' So I said to her, 'Well, I'm either going to fail and go back to the faculty, or I'm going to succeed and be a chancellor. I'm not sure I'm made of the political stuff to be a chancellor.' She said, 'You don't fool me for one minute!' (http://globetrotter.berkeley.edu/people/Ward/ward-con3.html).

Ward subsequently was appointed chancellor and later became president of the American Council on Education, and chancellor again at Wisconsin between 2011 and 2013. He showed a high level of effectiveness in these roles as a transformational leader, recognizing that support for an exemplary faculty was vital to the higher education mission. For example, disagreeing with a potent element of the Wisconsin legislature, he asserted in 2011: "My view is we can't be shy about competitive salaries." A faculty member responded to the press, "Chancellor Ward is saying exactly that at every public speech he gives to administrators, to faculty, to students and off-campus. . . . It seems to be his No. 1 priority, as well it should be" (http://host.madison.com/news/local/education/campus_connection/david-ward-says-uw-can-t-be-shy-about-competitive/article_b4221a84-f44f-11e0-ac44-001cc4c03286.html#ixzz38LnUlZ6d). Regardless, his impact went far beyond the University of Wisconsin, and his leadership in higher education has been pervasive. Leaders at all levels in academe today can learn much by emulating the manner in which David Ward has effectively practiced the craft of leadership.

11. Such third tier programs are far too frequently characterized by department heads, in addition to other faculty, that attempt to create a false image of excellence when this is not the case at all. I am one of many that encountered this sort of organizational behavior in an early graduate school experience. In programs where pretense is an accepted organizational cultural attribute, a shift to a new transformational leader as chair is especially important.

12. When I spoke with Reginald Golledge several years ago, he was very adamant about this one point, expressing dismay that many geography programs did little to enhance the employability of students in both undergraduate and graduate programs.

13. It is important to note here that practitioners of geography frequently find the research completed by those in other disciplines who share similar interests to be captivating; often bonds are formed, especially after the geographical nature of the common interest is explored. In fact, the Association of American Geographers offers an "Honorary Geographer" award annually as "a way of recognizing excellence in research, teaching, or writing on geographic topics by non-geographers." Gamma Theta Upsilon, the International Geographical Honor Society, also awards honorary membership to those who have "by exploration, research, lecturing, publishing teaching, or in any other way contributed to a marked advancement in any phase of geography." Certainly, these awards are ways that our discipline can boost its image, not only in academe, but in the public eye.

Yet, a number of geographers, both past and present, have refused to acknowledge that our discipline benefits from recognizing the inherently geographical work of those that do not have formal membership in our community. Perhaps it is a response to the contention that geography is not appropriate for university study that our field's representatives in academe often are wary of scholars in other disciplines. Or maybe some academic geographers simply seek to protect their own turf and wish not to share our discipline with those of a different lineage. It's also possible that personal animosities drive the makers of decisions, and it is emotion and not reason that results in missed opportunities.

One missed opportunity was a hundred years ago when U.S. President Woodrow Wilson, holding a Ph.D. in History and Political Science from Johns Hopkins University, among other things, was denied membership to the Association of American Geographers. Another missed opportunity was recent. A petition was submitted to the Executive Committee of Gamma Theta Upsilon in February 2014 for the International

Geographical Honor Society to award honorary membership to the noted anthropologist, conservationist, and author, Dr. Richard E. Leakey. Although Dr. Leakey more than met the requirements for the honorary membership, the Executive Committee refused to approve the petition. Thus, given the Wilson and Leakey cases, one is compelled to ask: How many more missed opportunities have there been in academic geography? How many more missed opportunities can geography afford?

14. Of course, other factors have played important parts in facilitating a department's success, such as program size and institutional support; however, the leadership behaviors of designated departmental academic leaders are crucial in affecting the department's place in the university, as well as in the discipline. This will become increasingly important as more doctoral programs emerge, greater parity among graduate programs becomes the norm, and competition for funds, faculty, and students becomes more and more fierce. Those characterized by leaders marked with a preponderance of transformational leadership behaviors, as well as faculty willing to work together toward a common goal, show a greater likelihood for achievement and prosperity.

Appendix A

Leadership in Academic Departments: A Review

The nature of academic departmental leadership comprises a small, but significant, component of the literature. Anderson (1977) has commented on the necessity of academic departments and Palm (1989) has commented on the need for program review to maintain departmental viability. Auclair (1990) conducted a thorough study of the historical development of leadership in academic departments. She noted that perhaps as much as 80 percent of the decisions made in higher education institutions are at the department level, and the chairperson might be the most important position in colleges and universities. This is a point that is iterated by Bennett (1988).

Hecht et al. (1999) maintained that because department chairs are the vital connection between the faculty and the administration, they must engage in effective communication and see that departmental objectives fall within the wider institutional mission. Thus, they offered a guidebook for the department chair that addresses a number of issues ranging from recruiting and evaluating faculty in the department to student advising, strategic planning, and relationships with the dean. The expectation for chairs is more now than it ever has been, for a higher level of accountability exists, and with that comes greater responsibility. In previous decades, the chairperson's principal role was, to large degree, devoted to serving as an advocate for the faculty in the department. That role has changed somewhat, for today the chair is tasked with understanding other departments' needs as well as the overall institutional mission and institutional relationships with agencies in the community. In essence, it is a leadership role that is plagued by complexities and ambiguities, but an imperative position in higher education.

Knight and Trowler (2000) indicated the imperative nature of sound leadership at the departmental level in improving effective teaching and learning. In a later study they (2001) also contended that chairpersons must learn to become effective leaders. Their list of important leadership traits and responsibilities (e.g., communicating a shared vision, open communication, motivation in teaching learning, and scholarship, mentoring and providing feedback) is consistent with what other students (e.g., Bass 1990; Arcuri 2002; Avolio 1994) have found to be inherent transformational leadership behaviors.

Martin et al. (1997) also investigated leadership in the academic department in relation to teaching. Their study was based on interviews with twenty-four department heads in the arts and sciences, science and engineering, law and commerce, and the health sciences. Some heads imposed decisions on their departments, while others engaged in collaborative decision-making with the faculty in their departments. Yet others accentuated the importance of the students' experience of studying, and these chairs sought to emphasize prospects for faculty to develop so that students' opportunities for development would be enhanced.

Also with reference to academic departments, House and Podsakoff (1994) embraced the concept of transformational leadership as effective leadership; they identified three important impacts, which create positive changes in followers: 1) Commitment to the leader's vision, 2) Disregard of self-interest for the sake of the team, and 3) Self-sacrificial behavior ("organizational citizenship") for the achievement of the collective vision. They argued that outstanding leaders create these effects upon their followers; bringing about significant changes in their organizations requires an articulation of their vision, passion, and self-sacrifice, as well as persistence in their efforts. The leader's image is considered highly important, as is the ability to instill within the followers a sense of self-confidence and high expectations. Mediocrity is not tolerated, which is consistent with other studies.

In essence, although some profound research in this area has been completed, study of academic departments and their leaders is lacking. Given the high level of decision-making made at the departmental level, more complete research on the nature of leadership in academic departments is an essentiality.

Appendix B

Sources on the History of Geography

In reference to the discipline's history, *All Possible Worlds* (Martin 2005), currently in its fourth edition, ranks as the authoritative record, Geoffrey Martin continues to be considered the leading specialist in the history of geographic thought worldwide, and his scholarship with regard to the development of the discipline has earned special praise from luminaries in the field (Dow 2005). Martin scoured the archives in universities across the United States in an attempt to unravel the story of the discipline's development in memoranda, correspondence, unpublished manuscript, and published research. His identification of marked paradigms in the discipline's history has gained wide acceptance (1985), and his research on the discipline's early years is noteworthy (1981; 1997).[1]

It must be noted that Martin's works on Mark Jefferson (1968), Isaiah Bowman (1973), and Ellsworth Huntington (1980) came as a result of substantial archival research employing Collingwood's (1946) thesis, which lend insight into the lives and thoughts of important figures that influenced the development of geography. In contrast to his prosopographies, a number of their students have offered biographical pieces of notable geographers. Arguably, no more has been written about any other geographer than Carl Sauer. Sauer's students and aficionados have surveyed elements of his writings and his life, offering tributes (e.g., Kenzer 1985; Kenzer 1987; Leighly 1963), commentaries (e.g., Denevan and Mathewson 2009; Entrikin 1984; Harris 2002; Leighly 1979; Mathewson and Kenzer 2003; Williams 1983), and personal accounts (e.g., Bruman 1996; Mikesell 1987; West 2004). Moreover, the *Geographers on Film* (GOF) series, founded by Maynard Weston Dow in 1970 with an interview of Sauer on the Berkeley campus, has proven to be particularly insightful. Subsequently, a number of Sauer's disciples offered reminiscences of their mentor (e.g., Dow 1970; 1971; 1973; 1976), which have since been transcribed.

Selected GOF interviews also lend great insight into the demise of the geography program at Harvard: Illick (Dow 1985), James (Dow 1984), Nash (Dow 1977), and Ullman (Dow 1972). These dialogues reveal much about the nature of the program's leadership under Derwent Whittlesey, which are complemented by Smith's (1987; 2003) written work on the subject as well. Smith's research, in addition to Martin's (1973; 1988) and

Glick's (1988) articles, document not only Isaiah Bowman's role in the Harvard debacle, but also Bowman's influence as a leader in the founding of a geography program at Johns Hopkins. Further, scholarship on academic geography's development is included in Blouet's (1981) anthology, *The Origins of Academic Geography in the United States*, James and Martin's (1979) *The Association of American Geographers: The first seventy-five years, 1904–1979*, and a number of articles by several students including: Dunbar (1986), Groop & Schaetzl (1997), James (1971), and Wilbanks (2004).

Essays incorporating reflections on practicing the discipline, as well as analyses of individual programs have also provided context. For example Hart (1979) reminisced about the status of the field at mid-century, and Kohn (1970) offered a perspective on the academic discipline in the 1960s. Barnes (2006) commented on the wartime service of academic geographers in the 1940s, which influenced their leadership roles upon returning to academe. Osofsky (2007) examined the decline of Yale's geography program within the context of the national scene and the development of its law school.

Also, studies devoted to the history and development of individual departments make the field rich for examining its past. The evolution of the geography program at Berkeley, for example, has been the subject of several articles; John Leighly (1979) wrote on graduate studies in the department during Sauer's early years, while James Parsons (1979) wrote about the program during Sauer's later years, and Bret Wallach (1998) offered a glimpse of the program during the post-Sauer years under Parsons. Kenzer (1985; 1987), Denevan and Mathewson (2009), and Mathewson and Kenzer (2003) have provided anthologies that offer in total a collection of valuable glimpses concerning the transformational leadership attributes of Carl Sauer, who presided as chair of the Berkeley department for three decades.

Though writings on other leaders in the field are not nearly as voluminous, they are substantive. For example, Golledge and Tobler's (1991) obituary of their colleague and Clarke and Baumgart's (2004) history of the Santa Barbara department shed much light on the leadership attributes of David Simonett, who spearheaded not only the creation of a new type of geography program, but a new paradigm in the history of geographic thought. Simonett's role as a transformational leader also is documented in Morain's (2006) memorial address, as well as the autobiographical pieces by Golledge and Tobler (Gould and Pitts 2002).

Elsewhere in California, Dunbar's (1981) brief synopsis of UCLA's geography department before the Second World War documents an important period in its development. Joseph Spencer (1979) offered his recollections of studying and teaching at both UCLA and Berkeley, and Nelson's (1985) tribute to Spencer offers insight as well. Reflections on

Jonathan Sauer are offered by former graduate students (Brothers et al. 2009) who discuss his inspirational character.

The first geography graduate program founded in the United States at the University of Chicago has been the subject of much scholarship; Visher (1953) discussed the influence of Rollin Salisbury, the eminent geomorphologist who oversaw its early development, as well as a number of later leaders. Pattison (1981) assessed Salisbury's contributions to the discipline as well as his leadership, and Harris (1979) provided a detailed account of the departmental leadership during the 1930s and 1940s under Harlan Barrows and Charles Colby. Harper (2005), Mikesell (2004), and Ullman in a GOF interview with Dow (1971) comment on the leadership attributes of Colby as well, while Rose (1964), Hartshorne (1964), and Mayer (1979) discuss other aspects of the Chicago department that related to its leaders. Berry and Dahmann's (2001) tribute to Wheatley is also an important contribution.

The University of Wisconsin's (Madison) departmental history has been documented in Glenn Trewartha's work (1978; 1979), as well as a number of tributes to luminaries that served in the department as faculty and passed on. Included among them are: Vernor Finch (Hartshorne 1961), Richard Hartshorne (Martin 1994), Glenn Trewartha (Hartshorne and Borchert 1988), Andrew Clark (Ward 1977), Arthur Robinson (Morrsion 2008), and J. Russell Whitaker (Martin 2003). Arthur Robinson's (1979) work on the history of cartography at Madison is especially important, and Rugg's (1981) contribution is useful for measuring the regional impact of the institution's alumni throughout academe. Moreover, Dow's interviews with Hartshorne (1984) and Robinson (1984) provide solid insight as well as Butzer's frank commentary on departmental politics (Gould and Pitts 2002).

Leadership in geography at the University of Michigan is assessed by examining a number of works; included among them are commentaries on Carl Sauer (Leighly 1976; Parsons 1976) and a GOF interview (Dow 1976) with Fred Kniffen, one of Sauer's early students. Preston James (Martin 1988) commented on his Michigan experience, as did Waldo Tobler and Gunnar Olssen (Gould and Pitts 2002).

Additionally, although Minnesota's department lacks a written history, tributes to Borchert (Adams 2007) and Cotton Mather (Karan 2003) provide insight to leadership in the program during the 1950s and 1960s, as does the interview of Mather by Clarke Chambers (1995). Fred Lukermann discussed the nature of the department as both a graduate student and faculty member; his account of the transformation of the department, which was discussed with Dow (1971) in a GOF interview provides valuable insight concerning the program after the suicide of eminent historical geographer Ralph Brown; John Borchert recounted the leadership of Jan Broek and the necessity for change (Dow 1975). Martin's (1994) memorial to Hartshorne, also lends insight to the character of the department before the Second World War under the authoritarian leadership of Darrell Davis.

A departmental history written by Guy Harold Smith and updated by Ned Taaffe (n.d.) provides a rich context for understanding the leadership and conditions faced by faculty and students at Ohio State University. Short biographies of Roderick Peattie (Smith 1957) and Ned Taaffe (Gauthier 2002) are supplemented by Arthur Robinson's GOF interview (Dow 1984), in which he recounted his experiences as a graduate student.

E. Willard Miller, the longtime chair at Penn State, authored a history of the department (1992). Larry Brown's (2004) tribute incorporates input from former students and colleagues of Miller, which creates a perspective on the leadership of the department. GOF interviews with former graduate students and faculty members, including Ron Abler (1985), John Adams (1983), and Allan Pred (1980), also provide insight, as does the obituary of longtime faculty member Peter Gould (Haggett 2003).

Surprisingly, no comprehensive account of the Syracuse program has been written, and here short biographies of George Cressey (Herman 1965; James 1964), Preston James (Martin 1988), and Donald Meinig (1992) must be relied upon, as well as Dow's GOF interviews with Preston James (1970, 1984) and J. Rowland Illick (1985). Despite the limited scholarship, these accounts provide valuable insight with regard to the transformational leadership practiced by Cressey and the mixture of leadership attributes that characterized James, both luminaries in their own right.

A number of articles have been written on Clark University's geography program. William Koelsch's four pieces (1980; 1987; 1988; 2002) concentrated on the first half of the twentieth century. Janice Monk (1998) examined female geography graduate students during the same period, Walter Ristow (1983) discussed field studies at the university in the 1930s, and Merle Prunty (1979) recounted his experiences as a graduate student in the 1940s. Martin (1988; 1998; 2005) touched upon Clark geography as well.

No departmental history of Johns Hopkins University has been published; yet it is an institution that created a significant impact on academic geography in America through national press reports in the period between the mid-1930s to the mid-1950s. The institution's president, Isaiah Bowman, graced the cover of *Time* magazine in 1936 (Education 1936); he was not only a notable geographer, but also an advisor to two U.S. presidents. Martin's (1980) prosopography provides insight into Bowman's role in the development of the geography program. Additionally, a number of pieces relating to Owen Lattimore, the Hopkins professor unjustly accused of Communism during the McCarthy era, provide a view of leadership in the geography program during the 1940s and 1950s (Dunbar 2000; Kirwan 1989; Kahin 1989; Sabin 1997; Hooper 1988; Freeman 1982). Of particular importance are works by Harvey (1983) and

Newman (1983). Corey's (1995) tribute to Jean Gottman, the eminent urban geographer that spent much of the 1940s at the Baltimore institution, offers insight as well.

Harvard's leadership in geography left much to be desired after the retirement of William Morris Davis, the luminary largely responsible for the formalization of geography as an academic discipline in the United States. Martin (2005) has discussed the early years of the twentieth century, and a number of works devoted to an examination of the Harvard program in the 1940s have offered ample evidence of Derwent Whittlesey's laissez faire leadership and the situation that unfolded resulting in the program's demise (Martin 1980; Martin 1988; Augelli and Patton 1988; Burghardt 1988; Cohen 1988; Glick 1988; Smith 1987; Smith 1988; Smith 2003). These works offer reflections of Harvard alumni and analyses by historians of the discipline that together create a valuable account of geography's final days as a viable discipline in Cambridge. Additionally, Ackerman's (1957) tribute to Whittlesey and Illick's (Dow 1985) GOF interview lend a more favorable view of the geography faculty at Harvard than other students.

Joseph Velikonja's (1994) thoroughly researched study of the history of Washington's geography program is nicely complemented by William Garrison's (1979) reflections as a faculty member during the "Quantitative Revolution." Additionally, John Sherman's (1991) tribute to G. Donald Hudson, the departmental chair that sparked significant changes in the course of graduate studies, explains much about the importance of effective leadership in facilitating success. Reflections on their experiences as graduate students are offered by Brian Berry, Waldo Tobler, and Richard Morrill (Gould and Pitts 2002); William Garrison discusses his tenure as a faculty member in Seattle as well (Gould and Pitts 2002). Additionally, GOF interviews with Brian Berry (Dow 1971), William Bunge (Dow 1976), William Garrison (Dow 1972), G. Donald Hudson (Dow 1971), and Edward Ullman (Dow 1972) contribute much to an assessment of leadership at the University of Washington during the 1950s. Donald Meinig (1992) also offered a brief glimpse of his experiences as a graduate student before quantitative methods were embraced as a paradigm in the department.

At the University of Iowa, numeracy represented a type of geography unlike that practiced throughout most of academe in the 1950s as well. James Lindberg's (2006) comprehensive departmental history offers a thorough chronological account of the people and events that contributed to making the department a notable program in American geography. King's (1988) tribute to H.H. McCarty also offers a valuable perspective on the development of the program, as does McCarty's (1979) article. Golledge (Gould and Pitts 2002) comments on his experiences in the program as well; Bunge (1979) discusses the impact of Schaefer; McNulty

(1991) offers a tribute to Clyde Kohn. King (1979) reflects on his experience as a graduate student at Iowa as well.

Geography at Northwestern University is recorded in a detailed essay by Martin Kenzer (1983), in which he discusses the pivotal role played by G. Donald Hudson as a catalyst in the development of the program. Hudson's leadership here is also discussed in his obituary (Sherman 1991); his GOF interview provides insight as well (Dow 1971). Hudson was clearly a transformational leader who facilitated the teaching and research agendas of several faculty. John Hudson (1993) discussed the leadership of Clarence Jones, certainly one of the best known of Northwestern geographers, and Gauthier's (2002) tribute to Ned Taaffe, a leader in the department during the 1950s, is a significant contribution as well. Taaffe's (1979) reflection offers valuable insight, as does Gould (1979), from a graduate student's perspective, and Espenshade's (Dow 1972) views are important as well.

Study of the history of leadership in second tier departments offers perspective on the challenges, conflicts, and successes of the discipline within a broader context. For example, Brett Wallach's (1999) frank assessment of departmental politics and problems associated with a succession of chairs at the University of Oklahoma is a cogent piece. Knapp's (1998) historical analysis of the Department of Geography at the University of Texas is a valuable contribution, as is Kimber and Hugill's (2000) examination of the department at Texas A&M. Nebraska's history has been documented by Hewes (1983), and his contributions as a chair are documented in an obituary (Wishart 2000).

In certain cases, though a comprehensive departmental historical analysis may be absent, tributes and obituaries devoted to significant leaders that served as chairmen are especially important. For example, Walter Kollmorgen's obituary (Shortridge et al. 2008), GOF interview (Dow 1970), and personal reflection (Kollmorgen 1979) offer ample evidence to support the contention that he was a transformational leader.

At Louisiana State University (LSU), Fred Kniffen, one of Carl Sauer's earliest students, established leadership over the geography program in Baton Rouge for many years. His transformational leadership character is evidenced in his GOF interview (Dow 1976), obituary (Walker and Richardson 1994), and the LSU departmental history authored by Mathewson and Shoemaker (1997). Elsewhere in the South, leadership at the University of Kentucky is examined in its departmental history (Karan and Brunn 2004), and Merle Prunty's influence at the University of Georgia is reflected in Hart's commentary (2004). Today, these departments might be considered at the top of the second tier, if not beyond.

Other relatively strong second tier departments include those at the University of Illinois, University of Oregon, University of Colorado, Arizona State University, University of Arizona, SUNY Buffalo, and Michigan State University. Departmental histories of the Arizona programs

(Comeaux 1987; Pederson 2007) provide much detail, and the Illinois history (Fellmann 1974) offers a solid chronology. Historical assessments of Colorado, Buffalo, and Michigan State remain to be written, and they are much needed. Moreover, though a departmental history of the University of Oregon has been written (Price 1990), a more critical assessment and an update is needed. At Oregon, Alexander Murphy's role as both a formal and informal leader, has been significant, and it is only a matter of time before a student of the history of geographic thought examines his contributions to the discipline, not only in terms of scholarship and education, but also in terms of leadership—and that might very well be the more important role.

There are others scattered across the country; for example, Duane Nellis, who sparked the development of a doctoral program at Kansas State, and went on to serve in administrative leadership positions there, and at West Virginia University as well, before taking the presidency of the University of Idaho is another.[2] And Stan Brunn, who led the department at the University of Kentucky to high levels of achievement and prosperity, is another. Certainly, leaders elsewhere also show promise. The point is that when it's all said and done, the good leaders tend to carry transformational leadership attributes—and these are behaviors that should be emulated for success.

Having said this, much can be gained from gathering the perspectives of academic geographers that have served in leadership roles. Unfortunately, with the relatively recent passing of Dow, the geographic discipline has lost its oral historian extraordinaire. Direly needed is continuation of *Geographers on Film* in some form, so that geographers can move beyond simply enjoying "dabbling" in the past and learn from their colleagues and leaders—not only in matters geographic, but also with regard to leadership strategies that will move the discipline forward.

NOTES

1. The work of Thomas Kuhn (1962) is particularly relevant, for the noted physicist commented on the development of scientific knowledge; paradigms, prevalent ways of thinking in individual academic disciplines, often dictated the content of disciplines and their subsequent trajectories. Kuhn suggested that the greater part of scientific development occurs not in a linear manner, but in revolutionary paradigm shifts. In the case of geography, the Kuhnian paradigm model has gained acceptance in examination of the history of geographic thought. Further, discussion of academic geography's origins in higher education is in order, for the leaders and programs that fostered the discipline's growth in its early days set the stage for later leaders to build upon the momentum of success—or conversely, collapse in despair.
2. Nellis recently took over as the president of Texas Tech.

References

Ackerman, E.A. (1957). Obituary: Derwent Stainthorpe Whittlesey. *Geographical review*, 47 (3), 443–445.

Adams, J.S. (2007). John Robert Borchert, 1918–2001. *Annals of the Association of American Geographers*, 97 (3), 641–648.

Aitken, S., and Valentine, G. (Eds.). (2006). *Approaches to human geography*. Thousand Oaks, CA: Sage.

Andersen, K. J. (1977). In defense of departments. In D. E. McHenry (Ed.), *Academic departments: Problems, variations, and alternatives* (pp. 1–11). San Francisco: Jossey-Bass, Inc.

Arcuri, N. (2002). *Faculty and chair perspectives on leadership and its impact on departmental outcomes*. Unpublished master's thesis, McGill University.

Aschmann, H. (1972). George McCutcheon McBride, 1876–1971. *Annals of the Association of American Geographers*, 62 (4), 685–688.

Association of American Geographers. (2010). *Guide to geography programs in the Americas, 2009–2010*. Washington, DC: Association of American Geographers.

Auclair, L.A. (1990). *Historical development of departmental leadership in American Higher education*. Unpublished doctoral dissertation, Boston College.

Augelli, J.P., and Patton, D.J. (1988). On "Academic War over the Field of Geography." *Annals of the Association of American Geographers*, 78 (1), 145–147.

Austin, A.E. (1996). Institutional and departmental cultures: The relationship between teaching and research. In J. M. Braxton (Ed.), *Faculty teaching and research: Is there a conflict* (pp. 57–66). San Francisco: Jossey-Bass.

Avolio, B.J. (1994). The "natural": Some antecedents to transformational leadership. *International journal of public administration*, 17, 1559–1581.

Avolio, B.J., and Bass, B.M. (1995). Individual consideration of analysis: A multi-level framework for examining the diffusion of transformational leadership. *The leadership quarterly*, 6 (2), 199–218.

Avolio, B.J., Bass, B.M., and Jung, D.I. (1999). Re-examining the components of transformational and transactional leadership using the multifactor leadership questionnaire. *Journal of occupational and organizational psychology*, 72, 441–462.

Avolio, B.J., Zhu, W., Koh, W, and Bhatia, P. (2004). Transformational leadership and organizational commitment: Mediating role of psychological empowerment and moderating role of structural distance. *Journal of organizational behavior*, 25 (8), 951–968.

Barbuto, J.E. (2005). Motivation and transactional, charismatic, and transformational leadership: A test of antecedents. *Journal of leadership and organizational studies*, 11 (4), 26–40.

Barnes, T.J. (2006). Geographical intelligence: American geographers and research and analysis in the Office of Strategic Services 1941–1945. *Journal of historical geography*, 32, 149–168.

Basham, L.M. (2010). *Presidents as transactional leaders in higher education*. Unpublished doctoral dissertation, Texas A & M University, Commerce.

Bass, B.M. (1985). *Leadership and performance beyond expectations*. New York: The Free Press.

Bass, B.M. (1990). From transactional to transformational leadership: Learning to share the vision. *Organizational dynamics*, 18 (3), 19–31.

Bass, B.M., and Avolio, B.J. (1990). The implications of transactional and transformational leadership for individual, team, and organizational development. In W.A. Passmore and R.W. Woodman (Eds.), *Research in organizational change and development* (pp. 231–272), Greenwich, CT: JAI Press Inc.

Bass, B.M., and Avolio, B.J. (1994). *Improving organizational effectiveness through transformational leadership*. Thousand Oaks, CA: Sage.

Bass, B.M., and Steidlmeier, P. (1999). Ethics, character, and authentic transformational leadership behavior. *Leadership quarterly*, 10 (2), 181–217.

Baumgart, S. (2004). History of the UCSB Geography Department. Retrieved November 24, 2004 from http://www.geog.ucsb.edu/more/history.

Beckinsale, R.P. (1981). W.M. Davis and American geography. In B.W. Blouet (Ed.), *The origins of academic geography in the United States* (pp. 107–122). Hamden, CT: Archon Books.

Beers, C. (1996). John C. Sherman: UW professor who shared his love of geography. *Seattle Times* (28 October).

Bennett, J.B. (1988). Department chairs: Leadership in the trenches. In M.F. Green (Ed.), *Leaders for a new era* (pp. 57–73). New York: MacMillan.

Berenson, B. (2008). Graduate education in the United States. In W. Smith and T. Bender (Eds.), *American higher education transformed, 1940-2005: Documenting the national discourse* (pp. 205–208). Baltimore: The Johns Hopkins University Press.

Berry, B.J.L. (2006). *Nihil sine labore*. Baltimore: Gateway Press.

Berry, B.J.L., and Dahmann, D.C. (2001). In memoriam: Paul Wheatley, 1921–1999. *Annals of the* Association *of American Geographers,* 91 (4), 734–747.

Berry, B.J.L., and Wheeler, J. (Eds.). (2005). *Urban geography in America, 1950–2000: Paradigms and personalities.* New York: Routledge.

Blouet, B.W., and Stitcher, T.L. (1981). Survey of early geography teaching in state universities and Land Grant institutions. In B.W. Blouet (Ed.), *The origins of academic geography in the United States* (pp. 327–342). Hamden, CT: Archon Books.

Bogdan, R.C., and Biklen, S.K. (2003). *Qualitative research for education: An introduction to theory and methods.* Boston: Allyn and Bacon.

Bono, J.E., and Judge, T.A. (2003). Self-concordance at work: Toward understanding the motivational effects of transformational leaders. *The Academy of Management journal*, 46 (5), 554–571.

Breit, H. (1950). Talk with Mr. Lattimore. *New York Times* (13 August).

Brooks, C.F. (1932). Robert DeCourcy Ward. *Geographical review,* 22 (1), 161.

Brooks, C.F. (1932). Robert DeCourcy Ward, Climatologist. *Annals of the Association of American Geographers*, 22 (1), 33–43.

Brothers, T.S., Fredrich, B., Gade, D.W., and Kimber, C.T. (2009). Jonathan D. Sauer, 1918–2008: Reflections on his life and work on Latin America and beyond. *Journal of Latin American geography*, 8 (1), 168–180.

Brown, L.A. (2004). In memoriam: E. Willard Miller, 1915–2002. *Annals of the Association of American Geographers,* 94 (3), 662–670.

Brown, S.E. (1978). Guy-Harold Smith, 1895–1976. *Annals of the Association of American Geographers*, 68 (1), 115–118.

Brown, W.R. (1982). *Academic politics*. University, Alabama: University of Alabama Press.

Browning, C. (1982). *Conversations with geographers: career pathways and research styles.* Chapel Hill, NC: University of North Carolina Department of Geography.

Bruman, H.J. (1996). Recollections of Carl Sauer and research in Latin America. *Geographical review,* 86 (3), 370–376.

Bunge, W. (1979). Fred K. Schaefer and the science of geography. *Annals of the Association of American Geographers*, 69 (1), 128–132.

Burghardt, A.P. (1988). On "Academic War over the Field of Geography," The elimination of geography at Harvard, 1947–1951. *Annals of the Association of American Geographers,*78 (1), 144.

Burns, J. M. (1978). *Leadership*. New York: Perennial.

Bushong, A.D. (1981). Geographers and their mentors: A genealogical view of American academic geography. In B.W. Blouet (Ed.), *The origins of academic geography in the United States* (pp. 193–219). Hamden, CT: Archon Books.

Carter, G.F. (1950). Isaiah Bowman, 1878–1950. *Annals of the Association of American Geographers,* 40 (4), 335–350.

Castree, N. (Ed.). (2006). Research assessment and the production of geographic knowledge. *Progress in Human Geography,* 30 (6), 746–782.

Cathcart, R.D. (2009). *Leadership in athletic coaches.* Unpublished doctoral dissertation, Indiana University of Pennsylvania.

Chambers, C.A. (1995). Unpublished interview with Eugene Cotton Mather.

Clarke, K. C., and Baumgart, S. R. (2004). The Department of Geography at the University of California, Santa Barbara: History, curriculum, and pedagogy. *Yearbook of the Association of Pacific Coast Geographers, 66,* 95–113.

Cohen, S.B. (1988). Reflections on the elimination of geography at Harvard, 1947–1951. *Annals of the Association of American Geographers,* 78 (1), 148–151.

Colby, C.C. (1936). Changing currents of geographic thought in America. *Annals of the Association of American Geographers,* 26 (1), 1–37.

Colby, C.C., and White, G.F. (1961). Harlan H. Barrows, 1877–1960. *Annals of the Association of American Geographers,* 51 (4), 395–400.

Collingwood, R.G. (1946). *The idea of history.* New York: Oxford University Press.

Comeaux, M.L. (1987). *101 years of geography at Arizona State University.* Department of Geography Publication Series 2. Tempe: Arizona State University.

Cressey, G.B. (1949). Wallace W. Atwood, 1872–1949. *Annals of the Association of American Geographers,* 39 (4), 296–306.

Czech, K. (2007). *Communication and leadership: Faculty perceptions of the department chair.* Unpublished doctoral dissertation, University of San Diego.

Davis, W.M. (1932). The college life of Robert DeCourcy Ward. *Annals of the Association of American Geographers,* 22 (1), 28–32.

Denevan, W.M. (1987). Observations on trade and gold in the early Spanish Main by Carl Sauer. In M.S. Kenzer (Ed.), *Carl O. Sauer: A tribute* (pp. 164–174). Corvallis, OR: Oregon State University Press.

Denevan, W.M., and Mathewson, K. (Eds.). (2009). *Carl Sauer on culture and landscape: Readings and commentaries.* Baton Rouge, LA: Louisiana State University Press.

Den Hertog, D. N., Van Muijen, J. J., and Koopman, P. L. (1997). Transactional versus transformational leadership: Analysis of the MLQ. *Journal of Occupational and Organizational Psychology,* 70, 19–34.

DeVivo, M. S. (1987, April.). *The emergence and decline of physical geography in the United States.* Paper presented at Association of American Geographers Annual Meeting, Portland, OR.

DeVivo, M.S. (1988). *The evolution of basic physical geography, 1892–1967: An analysis of American physical geography textbooks.* Unpublished master's thesis. University of Tennessee, Knoxville.

DeVivo, M. S. (1990). Geography and geographers at the University of Tennessee, 1794–1989: The history of a collegiate geography department in Appalachia. *Proceedings of the Fifth Conference on Appalachian Geography, 5,* 74–89.

DeVivo, M.S. (2006, October). *The emergence and decline of geography departments in the United States.* Paper presented at National Council of Geographic Education Annual Meeting, Lake Tahoe, NV.

DeVivo, M.S. (2011). A Tale of Two Departments: Success and Failure in Higher Education Leadership in American Geography at Mid-Century. *International Conference on Social Science and Humanity Proceedings,* VI-243-VI-247.

Dobson, J.E., Trimble, S.W., Dobson, J.R., and Friend, D.A. (1999). In memoriam: Kirk H. Stone, 1914–1997. *Annals of the Association of American Geographers,* 89 (3), 535–548.

Doctoral Programs by the Numbers. (2010, 1 October). *Chronicle of higher education.* Retrieved 2 October 2010, from: http://0-chronicle.com.lib.grcc.edu/article/Tables-Doctoral-Programs-by/124789/

Dow, M.W. (1970). *Jan O. M. Broek interviewed by Preston E. James.* Online Geographers on Film transcription. Unpublished manuscript.

Dow, M.W. (1970). *George F. Carter interviewed by Preston E. James.* Online Geographers on Film transcription. Unpublished manuscript.

Dow, M.W. (1970). *Preston E. James interviewed by Maynard Weston Dow.* Online Geographers on Film transcription. Unpublished manuscript.

Dow, M.W. (1970). *Walter Kollmorgen interviewed by Preston E. James.* Online Geographers on Film transcription. Unpublished manuscript.

Dow, M.W. (1970). *John B. Leighly interviewed by Preston E. James.* Online Geographers on Film transcription. Unpublished manuscript.

Dow, M.W. (1970). *Carl O. Sauer interviewed by Preston E. James.* Online Geographers on Film transcription. Unpublished manuscript.

Dow, M.W. (1971). *Marvin W. Mikesell interviewed by John Fraser Hart.* Online Geographers on Film transcription. Unpublished manuscript.

Dow, M.W. (1971). *Fred Lukermann interviewed by Louis Seig.* Online Geographers on Film transcription. Unpublished manuscript.

Dow, M.W. (1972). *Edward Espenshade interviewed by Preston E. James.* Online Geographers on Film transcription. Unpublished manuscript.

Dow, M.W. (1972). *Edward L. Ullman interviewed by Preston E. James.* Online Geographers on Film transcription. Unpublished manuscript.

Dow, M.W. (1972). *Richard Hartshorne interviewed by Preston E. James.* Online Geographers on Film transcription. Unpublished manuscript.

Dow, M.W. (1973). *Hallock F. Raup interviewed by John Fraser Hart.* Online Geographers on Film transcription. Unpublished manuscript.

Dow, M.W. (1974). The oral history of geography. *The professional geographer*, 26, 430–435.

Dow, M.W. (1975). *John R. Borchert interviewed by Louis Seig.* Online Geographers on Film transcription. Unpublished manuscript.

Dow, M.W. (1976). *Fred B. Kniffen interviewed by William Haag.* Online Geographers on Film transcription. Unpublished manuscript.

Dow, M.W. (1977). *Peter Nash interviewed by Maynard Weston Dow.* Online Geographers on Film transcription. Unpublished manuscript.

Dow, M.W. (1980). *Allan Pred interviewed by Geoffrey J. Martin.* Online Geographers on Film transcription. Unpublished manuscript.

Dow, M.W. (1983). *John S. Adams interviewed by Maynard Weston Dow.* Online Geographers on Film transcription. Unpublished manuscript.

Dow, M.W. (1984). *Preston E. James interviewed by Maynard Weston Dow.* Online Geographers on Film transcription. Unpublished manuscript.

Dow, M.W. (1984). *Arthur H. Robinson interviewed by Maynard Weston Dow.* Online Geographers on Film transcription. Unpublished manuscript.

Dow, M.W. (1984). *Wilbur Zelinsky interviewed by Maynard Weston Dow.* Online Geographers on Film transcription. Unpublished manuscript.

Dow, M.W. (1984). *Gilbert F. White interviewed by Geoffrey J. Martin.* Online Geographers on Film transcription. Unpublished manuscript.

Dow, M.W. (1985). *Ronald F. Abler interviewed by Maynard Weston Dow.* Online Geographers on Film transcription. Unpublished manuscript.

Dow, M.W. (1985). *J. Rowland Illick interviewed by Maynard Weston Dow.* Online Geographers on Film transcription. Unpublished manuscript.

Dow, M.W. (1986). *Richard Hartshorne interviewed by Maynard Weston Dow.* Online Geographers on Film transcription. Unpublished manuscript.

Dow, M.W. (2005). All Possible Worlds: A History of Geographical Ideas. *Choice*, 43 (1), 160–161.

Dunbar, G.S. (1981). *Geography in the University of California (Berkeley and Los Angeles), 1868–1941*. Marina del Rey, CA: DeVorss.
Dunbar, G.S. (1986). Geography in the bellwether universities of the United States. *Area, 18(1)*, 25–33.
Dunbar, G.S. (2000). Owen Lattimore: 1900–1989. *Geographers: Biobibliographical studies*, 20, 24–42.
Eckel, P. (1998).Thinking differently about academic departments: The academic department as a team. In S.H. Frost (Ed.), *Using teams in higher education: Cultural foundations for productive change* (pp. 27–38). San Francisco: Jossey-Bass Publishers.
Education: Scholars without money. (1936). *Time*, 27, (12).
Eid, J., Johnsen, B.J., Brun, W, Laberg, J.C., and Nyhus, J. (2004). Situational awareness and transformational leadership in senior military leaders: An exploratory study. *Military psychology*, 16 (3), 203–209.
Entrikin, N. (1984). Carl O. Sauer: Philosopher in spite of himself. *Geographical review*, 74 (4), 389–408.
Erdman, H. (2002). *Transformational leadership congruence: A comparison between university and community college presidential leadership*. Unpublished doctoral dissertation, Texas A & M University, Commerce.
Felfe, J., Six, B., and Tartler, K. (2004). The impact of followers' motives, cognitive abilities, and personality on the perception of transformational leadership. *International journal of psychology*, 39 (5–6), 129.
Fellmann, J.D. (1974). *Geography at Illinois: The discipline and the department, 1867–1974*. Unpublished manuscript.
Figueroa, R.A. (2007). *Higher education chancellors in Puerto Rico: Strategies and activities implemented to develop transformational leadership in their institutions*. Unpublished doctoral dissertation, University of Puerto Rico, Rio Piedras.
Firestone, D.T. (2007). *A study of leadership behaviors among chairs in allied health programs*. Unpublished doctoral dissertation, St. John's University.
Footit, P.C. (2000). *Transformational leadership practices of physical education chairpersons in higher education*. Unpublished doctoral dissertation, Northern Illinois University.
Freeman, M.V. (1982). Abe Fortas: Man of courage. *The Yale law journal*, 91 (6), 1052–1060.
Gardner, H. (1995). *Leading minds*. New York: Basic Books.
Garrison, W.H. (1979). Playing with ideas. *Annals of the Association of American Geographers*, 69 (1), 118–120.
Gauthier, H.L. (2002). In memoriam: Edward "Ned" Taaffe, (1921–2001). *Annals of the Association of American Geographers*, 92 (3), 573–583.
Gibson, J.R. (Ed.) (1978). *European settlement and development in North America: Essays on geographical change in honour and memory of Andrew Hill Clark*. Toronto: University of Toronto, Department of Geography Research Publication no. 15.
Gilbert, P.K. (1997). *Transformational and transactional leadership by community college presidents*. Unpublished doctoral dissertation, University of Florida.
Gittens, B.E. (1996). *Perceptions of the applicability of transformational leadership behavior to the leader role of academic department chairs: A study of selected universities in Virginia*. Unpublished doctoral dissertation, George Washington University.
Glenn, D. (2010, 30 September). A critic sees deep problems in the doctoral rankings. *Chronicle of higher education*. Retrieved 2 October 2010 from: http://0-chronicle.com.lib.grcc.edu/article/A-Critic-Sees-Deep-Problems-in/124725/.
Glick, T.F. (1988). Before the revolution: Edward Ullman and the crisis of geography at Harvard. In J.E. Harmon and T.J. Rickart, *Geography in New England: A special publication of the New England/St. Lawrence Valley Geographical Society* (pp. 49–62).
Goldberger, M.L., Maher, B.A., and Flattau, P.E. (Eds.). (1995). *Research-doctorate programs in the United States: Continuity and change*. Washington, DC: National Academy Press.
Golledge, R., and Tobler, W. (1990). David Simonett, Ph.D. *Remote Sensing Specialty Group of the Association of American Geographers newsletter*.

Gould, P. (1979). Geography, 1957–1977: The Augean period. *Annals of the Association of American Geographers,* 79 (1), 139–151.
Gould, P, and Pitts, F.R. (Eds.) (2002). *Geographical voices: Fourteen autobiographical essays*. Syracuse: Syracuse University Press.
Griess, P.R. (1951). The Bolivian tin industry. *Economic geography,* 27 (3), 238–250.
Groop, R.E., and Schaetzl, R.J. (1997). Productivity profiles of PhD-granting geography departments in the United States: 1980–1994. *Professional geographer,* 49 (4), 451–464.
Grosso, F.A. (2008). *Motivating faculty through transformational leadership: A study of the relationship between presidential leadership behaviors and faculty behaviors*. Unpublished doctoral dissertation, Catholic University of America.
Haggett, P. (2003). In memoriam: Peter Robin Gould, 1932–2000. *Annals of the Association of American Geographers,* 93 (4), 925–934.
Hall, R.B. (1953). William Herbert Hobbs, 1864–1953. *Annals of the Association of American Geographers,* 43 (4), 284–288.
Harash, H.A. (2010). *An analysis of the relationship between the perceived leadership styles of educational leaders and the job satisfaction of faculty members who serve under them within community colleges.* Unpublished doctoral dissertation, Pepperdine University.
Harper, R.A. (2005). *A short history of geography at Southern Illinois University Carbondale, 1868–1967.* Unpublished manuscript.
Harris, C.D. (1966). Charles C. Colby, 1884–1965. *Annals of the Association of American Geographers,* 56 (2), 378–382.
Harris, C.D. (1977). Edward Louis Ullman, 1912–1976. *Annals of the Association of American Geographers,* 67 (4), 595–600.
Harris, C.D. (1979). Geography at Chicago in the 1930s and 1940s. *Annals of the Association of American Geographers, 69(1),* 21–32.
Harris, D.R. (2002). "The farther reaches of human time": Retrospect on Carl Sauer as prehistorian. *Geographical review,* 92 (4), 526–544.
Hart, J.F. (1979) The 1950s. *Annals of the Association of American Geographers,* 69 (1), 109–114.
Hart, J.F. (2004). Early days at Georgia. In J. Wheeler and S. Brunn (Eds.), *The role of the South in the making of American geography,* (pp. 218–225). Columbia, MD: Bellwether Press.
Hartshorne, R. (1939). The nature of geography: A critical survey of current thought in the light of the past. *Annals of the Association of American Geographers,* 29, 163–658.
Hartshorne, R. (1961). Vernor Clifford Finch, 1883–1959. *Annals of the Association of American Geographers,* 51 (3), 339–342.
Hartshorne, R. (1964). Robert S. Platt, 1891–1964. *Annals of the Association of American Geographers,* 54 (4), 630–637.
Hartshorne, R. (1981). William Morris Davis—The course of development of his concept of geography. In B. W. Blouet (Ed.), *The origins of academic geography in the United States* (pp. 139–149). Hamden, CT: Archon Books.
Hartshorne, R. and Borchert, J. (1988). In memoriam: Glenn T. Trewartha, 1896–1984. *Annals of the Association of American Geographers,* 78 (4), 728–735.
Harvey, D. (1983). Owen Lattimore: A memoire. *Antipode,* 15 (3), 3–11.
Hecht, I.W.D., Higgerson, M.L., Gmelch, Walter H., and Tucker, A. (1999). *The department chair as academic leader*. Phoenix, AZ: American Council on Education and The Oryx Press.
Herman, T. (1965), George Babcock Cressey, 1896–1963. *Annals of the Association of American Geographers,* 55 (2), 360–364.
Hewes, L. (1983). Geography at the University of Nebraska. *Great Plains-Rocky Mountains geographical journal,* 11.
Hilton, R. (1999). Harvard and Stanford. Retrieved December 7, 2004, from http://wais.stanford.edu/Geography/geography_harvardstanford.html.
Hinkin, T.R., and Schriesheim, C.A. (2008). A theoretical and empirical examination of the transactional and non-leadership dimensions of the multifactor leadership questionnaire (MLQ). *The leadership quarterly,* 19 (5), 501–513.

Hooper, P.F. (1988). The Institute of Pacific Relations and the origins of Asian and Pacific studies. *Pacific affairs*, 61 (1), 98–121.
Horwitz, T. (2012). Looking at the Battle of Gettysburg through Robert E. Lee's eyes. *Smithsonian*, 43, 88–93.
House, R.J., and Podsakoff, P.M. (1994). Leadership effectiveness: Past perspectives and future directions for research. In J. Greenburg (Ed.), *Organizational behavior: The state of the science* (pp. 45–82). Hillsdale, NJ: Erlbaum.
Hudson, J.C. (1993). In memoriam: Clarence Fielden Jones, 1893–1991. *Annals of the Association of American Geographers*, 83 (1), 167–172.
Jackson, R.L. (1999). *Transformational and transactional leadership in division administration at three institutions of higher education: An application of Bass and Avolio Multifactor Leadership Questionnaire*. Unpublished doctoral dissertation, University of Cincinnati.
James, P.E. (1935). *An outline of geography*. Boston: Ginn and Company.
James, P.E. (1964). George Babcock Cressey, 1896–1963. *Geographical review*, 54 (2), 254–257.
James, P.E. (1971). The significance of geography in American education. In D.W. Meinig (Ed.), *On geography: The selected writings of Preston James* (pp. 370–389). Syracuse, NY: Syracuse University Press.
James, P.E., and Martin, G.J. (1978). The *Association of American Geographers: The first seventy-five years, 1904–1979*. Washington, DC: Association of American Geographers.
Johnson, L. (1971). Raymond E. Murphy: An appreciation. *Economic geography* 47 (2), 97–100.
Jones, L.V., Gardner, L, and Coggeshall, P.E. (Eds.) (1982). *An assessment of research doctorate programs in the United States: Social and Behavioral Sciences*. Washington, DC: National Academy Press.
Jung, D.I., and Avolio, B.J. (2000). Opening the black box: An experimental investigation of the mediating effects of trust and value congruence on transformational and transactional leadership. *Journal of organizational behavior*, 21 (8), 949–964.
Kahin, G.M. (1989) Owen Lattimore (1900-1989). *The journal of Asian studies*, 48 (4), 945–946.
Karan, P.P. (2003). In memoriam: Cotton Mather, (1918–1999). *Annals of the Association of American Geographers*, 93 (2), 485–493.
Karan, P.P., and Brunn, S.D. (2004). Sixty years of geography at the University of Kentucky, 1944–2004. In J. Wheeler and S. Brunn (Eds.), *The role of the South in the making of American geography*, (pp. 166–172). Columbia, MD: Bellwether Press.
Keniston, H. (1959). *Graduate study in the arts and sciences at the University of Pennsylvania* Philadelphia: University of Pennsylvania Press.
Kenzer, M.S. (1983). The formation of an independent department of geography at Northwestern University: A chapter in the growth of American geography after World War II. *History of geography newsletter*, 3 (December), 30–37.
Kenzer, M.S. (1985). Milieu and the "intellectual landscape": Carl O. Sauer's undergraduate heritage. *Annals of the Association of American Geographers*, 75 (2), 258–270.
Kenzer, M.S. (1987). *Carl Sauer: A tribute*. Corvallis, OR: Oregon State University Press.
Kimber, C., and Hugill, P.J. (2000). Berkeley-on-the Brazos and other pipe dreams: History of the department of geography at Texas Aand M University. *Southwestern geographer*, 4, 94–120.
King, K.C. (1994). *Transformational leadership in School of Nursing deans: Perceived effectiveness, faculty satisfaction, and faculty willingness to put forth extra effort*. Unpublished doctoral dissertation, Spalding University.
King, L.J. (1979). Areal associations and regressions. *Annals of the Association of American Geographers*, 69 (1), 124–128.
Kirkman, M. (2004). *Leadership style of community college department chairs and the effect of faculty job satisfaction*. Unpublished doctoral dissertation, University of Missouri-Columbia.

Kirwan, L.P. (1989). Professor Owen Lattimore, 1900–1989. *The geographical journal,* 155 (3), 448.

Knapp, G. (1998). Geography at the University of Texas at Austin. *The Southwestern Geographer,* 2, 97–112.

Knight, P.T., and Trowler, P.R. (2000). Department-level cultures and the improvement of learning and teaching. *Studies in higher education,* 25 (1), 69–83.

Knight, P.T., and Trowler, P.R. (2001). *Departmental leadership in higher education*. Buckingham, UK: The Society for Research into Higher Education and Open University Press.

Koelsch, W.A. (1969). The historical geography of Harlan H. Barrows. *Annals of the Association of American Geographers,* 59 (4): 632–651.

Koelsch, W.A. (1980). Wallace Atwood's "Great Geographical Institute." *Annals of the Association of American Geographers,* 70 (4), 567–582.

Koelsch, W.A. (1987). *Clark University, 1887–1987: A narrative history.* Worcester, MA: Clark University Press.

Koelsch, W.A. (1988). Geography at Clark: The first fifty years, 1921–1971. In J.E. Harmon and T.J. Rickart (Eds.), *Geography in New England: A special publication of the New England/St. Lawrence Valley Geographical Society* (40–48).

Koelsch, W.A. (2002). G. Stanley Hall, child study, and the teaching of geography. *Journal of Geography* 101, 3–9.

Kohn, C.F. (1970). The 1960s: A decade of progress in geographical research and instruction. *Annals of the Association of American Geographers,* 60 (2), 211–219.

Kollmorgen, W.M. (1979). Kollmorgen as a bureaucrat. *Annals of the Association of American Geographers,* 69 (1), 77–89.

Kohn, C.F. (1970). The 1960s: A decade of progress in geographical research and instruction. *Annals of the Association of American Geographers,* 60 (2), 211–219.

Kuhn, T.S. (1962). The *structure of scientific revolutions*. Chicago: University of Chicago Press.

Kull, S.L. (2003). *The ameliorating effects of transformational leadership on resistance to change. Transformational leadership and conductivity for schematic conversion*. Unpublished doctoral dissertation, Regent University.

Lasswell, H.D. (2005) The policy orientation. In Stehr, N., and Grundmann (Eds.), *Knowledge: Critical Concepts* (pp. 40–55). New York, Routledge.

Lawson, V. (2004). Healthy departments, healthy discipline. *Association of American Geographers newsletter,* 39(8), 3.

Leighly, J. (1937). Some comments on contemporary geographic method. *Annals of the Association of American Geographers,* 27 (3), 125–141.

Leighly, J. (Ed.) (1963). *Land and life, a selection from the writings of Carl Ortwin Sauer.* Berkeley, CA: University of California Press.

Leighly, J.B. (1976). Carl Ortwin Sauer, 1889–1975. *Annals of the Association of American Geographers,* 66 (3), 337–348.

Leighly, J.B. (1979). Berkeley: Drifting into geography in the Twenties. *Annals of the Association of American Geographers,* 69 (1), 4–9.

Levine, M.F. (2000). *The importance of leadership: An investigation of presidential style at fifty national universities*. Unpublished doctoral dissertation, University of North Texas.

Lindberg, J. (2006). *History of the Geography Department, University of Iowa*. Unpublished manuscript.

Luton, B. (2010). *Transformational leadership and organizational commitment: A study of UNC system business school department chairs*. Unpublished doctoral dissertation, Capella University.

Martin, E., Prosser, M., Benjamin, J., Trigwell, K., and Ramsden, P. (1997). Heads' of academic departments' conceptions of leadership of teaching. In C. Rust and G. Gibbs (Eds.), *Improving student learning: Improving student learning through course design*. Oxford, UK: Oxford Centre for Staff and Learning Development.

Martin, G.J. (1968). *Mark Jefferson: Geographer*. Ypsilanti, Michigan: Eastern Michigan University Press.

Martin, G.J. (1973). *Ellsworth Huntington: His life and thought*. Hamden, CT: Archon Books.

Martin, G.J. (1980). *The life and thought of Isaiah Bowman*. Hamden, CT: Archon Books.

Martin, G.J. (1981). Ontography and Davisian physiography. In B.W. Blouet (Ed.), *The origins of academic geography in the United States* (pp. 279–289). Hamden, CT: Archon Books.

Martin, G.J. (1985). Paradigm change: A history of geography in the United States, 1892–1925. National *Geographic research* (Spring), 217–235.

Martin, G.J. (1987). Foreword. In M.S. Kenzer (Ed.), *Carl Sauer: A tribute* (pp. ix–xvi). Corvallis, OR: Oregon State University Press.

Martin, G.J. (1988). On Whittlesey, Bowman, and Harvard. *Annals of the Association of American Geographers*, 78 (1), 152–158.

Martin, G.J. (1988). In memoriam: Preston E. James, 1899–1986. *Annals of the Association of American Geographers*, 78 (1), 164–175.

Martin, G.J. (1994). In memoriam: Richard Hartshorne, 1899–1992. *Annals of the Association of American Geographers*, 84 (3), 480–492.

Martin, G. J. (1997). Geography in New England prior to the "New England Geographical Conference of 1922." *Proceedings: New England-St. Lawrence Valley Geographical Society*, *27*, 14–22.

Martin, G.J. (1998). The emergence and development of geographic thought in New England. *Economic geography* (extra issue), 1–13.

Martin, G.J. (2003). In memoriam: Joe Russell Whitaker (1900–2000). *Annals of the Association of American Geographers*, 93 (1), 223–228.

Martin, G.J. (2003). From the cycle of erosion to the "Morphology of Landscape": Or some thought concerning geography as it was in the early years of Carl Sauer. In K. Mathewson and M.S. Kenzer (Eds.). *Culture, land, and legacy: Perspectives on Carl O. Sauer and Berkeley School geography*. Baton Rouge, LA: Geoscience Publication.

Martin, G.J. (2005). *All possible worlds: A history of geographical ideas* (4th ed.). New York: Oxford University Press.

Massaro, A.V. (2007). *Exploring the learning paths of academic department chairs*. Unpublished doctoral dissertation, The Ohio State University.

Mathewson, K., and Kenzer, M.S. (Eds.). (2003). *Culture, land and legacy: Carl Sauer and Berkeley School geography*. Baton Rouge, LA: Geoscience Publication, Louisiana State University.

Mathewson, K., and Shoemaker, V. J. (1997). Louisiana State University's Department of Geography and Anthropology: A selective history. *The Southwestern Geographer*: 76–101.

Mayall, W.T. (2008). *Assessing National Defense University presidents against the intellectual stimulation component of Bass's transformational leadership model*. Unpublished doctoral dissertation, George Washington University.

Mayer, H.M. (1979). Urban geography and Chicago in retrospect. *Annals of the Association of American Geographers*, 69 (1), 114–118.

McCarty, H.H. (1979). The Cornbelt connection: Geography at Iowa. *Annals of the Association of American Geographers*, 69 (1), 121–124.

McMaster, R., and McMaster, S. (2002). A history of twentieth-century American academic cartography. *Cartography and Geographic Information Sciences*, 29 (3), 305–321.

McNulty, M.L. (1991). In memoriam: Clyde F. Kohn, 1911–1989. *Annals of the Association of American Geographers*, 81 (4), 697–700.

Meinig, D.W. (1953). *The Walla Walla country: 1805-1910: A century of Man and the land*. Unpublished doctoral dissertation, University of Washington.

Meinig, D.W. (1989). The historical geography imperative, *Annals of the Association of American Geographers*, 79 (1), 79–87.

Meinig, D.W. (1992). A life of learning. *American Council of Learned Societies Occasional Paper 19*.

Mikesell, M.W. (1987). Sauer and "Sauerology": A student's perspective. In M.S. Kenzer (Ed.), *Carl O. Sauer: A tribute* (pp. 144–150). Corvallis, OR: Oregon State University Press.
Mikesell, M.W. (2004). In memoriam: Chauncy Dennison Harris, (1914–2003). *Annals of the Association of American Geographers*, 94 (4), 982–991.
Miller, E.W. (1992). *The College of Earth and Mineral Sciences at Penn State*. University Park, PA: Pennsylvania State University Press.
Mind Garden. (2003) MLQ for transformational measurement. Retrieved from: http://www.mindgarden.com/
Monk, J. (1998). The women were always welcome at Clark. *Economic geography*, 78 (extra issue), 14–30.
Monk, J. (2004). Women, gender, and the histories of American geography. *Annals of the Association of American Geographers*, 94 (1), 1–22.
Morain, S.A. (2006). David Stanley Simonett, 1926–1990. *Photogrammetric engineering and remote sensing*, 72 (7), 763–765.
Murphy, A.B. (2007). Geography's place in higher education in the United States. *Journal of geography in higher education*, 31 (1), 121–142.
Murphy, A.B. (2008). The extraordinary career of Gilbert White. *Progress in human geography*, 32 (3), 445–461.
Murphy, R.E. (1932). Geography and Mineral Economics at Penn State. *Mineral Industries* 1(6), 1.
Murphy, R.E. (1979). Economic Geography and Clark University. *Annals of the Association of American Geographers, 69(1)*, 39–42.
Nelson, H.J. (1985). In memoriam: J.E. Spencer, 1907–1984. *Annals of the Association of American Geographers*, 75 (4), 595–603.
Newman, R.P. (1983). Lattimore and his enemies. *Antipode*, 15 (3), 12–26.
Northouse, P.G. (1997). *Leadership: Theory and practice.* Thousand Oaks, CA: Sage.
Obituary: Ralph Hall Brown. (1948). *Geographical review*, 38 (3), 506–507.
Ogle, E.H. (2000). *The art of being human: A case study of transformational leadership.* Unpublished doctoral dissertation, University of California, Los Angeles.
Ormeling, F.J. (1975). Obituary: Profile of the geographer Jan Broek. *Tijdschrift voor economische en sociale geografie*, 66 (1), 3–5.
Osofsky, H. M. (2007). A law and geography perspective on the New Haven School. *The Yale journal of international law*, 32 (42), 421–453.
Palm, R. (1989). The geographer as administrator: Perspectives on survival for geography departments. In M. S. Kenzer (Ed.), *On becoming a professional geographer* (pp. 203–211). Columbus, OH: Merrill Publishing Company.
Para, D.J. (2010). *Leadership qualities desired for college presidents as perceived by college trustees*. Unpublished doctoral dissertation, Capella University.
Parry, K.W., and Proctor-Thomson, S.B. (2002). Perceived integrity of transformational leaders in organizational settings. *Journal of business ethics*, 35, 75–96.
Parsons, J.J. (1976). Carl Ortwin Sauer, 1889–1975. *Geographical review*, 66 (1), 83–89.
Parsons, J.J. (1979). The later Sauer years. *Annals of the Association of American Geographers*, 69 (1), 9–18.
Parsons, J.J. (1986). Carl Sauer's vision of an institute of Latin American studies. *Geographical review*, 86 (3), 377–384.
Parsons, J.J., Hooson, D., Oberlander, T., and Stanislawski, D. (1986). John Leighly, Geography, Berkeley. In D. Krogh (Ed.), *University of California: In memoriam, 1986*, (pp. 174–177). Oakland, CA: Academic Senate, University of California.
Pattison, W.D. (1981). Rollin Salisbury and the establishment of geography at the University of Chicago. In B.W. Blouet (Ed.), *The origins of academic geography in the United States* (pp. 151–163). Hamden, CT: Archon Books.
Paul Dean Simkins. (2010). Obituary. *Centre Daily Times* (13 February).
Pederson, L.R. (2002). *Geography and regional development at the University of Arizona: A history*. Unpublished manuscript.

Platt, R.S. (1957). A note on Rollin Salisbury. *Annals of the Association of American Geographers*, 47 (3), p. 276.

Post, J.M. (2004). *Leaders and their followers in a dangerous world: The psychology of political behavior*. Ithaca, NY: Cornell University Press.

Powers, D. (1979). *Syracuse University Hall of Languages rededication*. Syracuse: Syracuse University.

Price, E.T. (1990). Geography at the University of Oregon. *Association of Pacific Coast Geographers Yearbook*, 52, 140–152.

Prunty, M.C. (1979). Clark in the early 1940s. *Annals of the Association of American Geographers*, 69(1), 42–45.

Ristow, W.W. (1983). Three months in the field: September 12 to December 12, 1934. *History of geography newsletter*, 3, 21–29.

Robinson, A.H. (1979). Geography and cartography then and now. *Annals of the Association of American Geographers*, 69 (1), 97–102.

Roose, K.D., and Andersen, C.J. (1970). *A rating of graduate programs*. Washington, DC: American Council on Education.

Rose, J.K. (1964). Griffith Taylor, 1880–1963. *Annals of the Association of American Geographers*, 54 (4), 622–629.

Rostlund, E. (1956). Twentieth-century magic. *Landscape*, 5 (3), 23–26.

Roueche, J.E., Baker III, G.A., and Rose, R.R. (1989). *Shared vision: Transformational leadership in American community colleges*. Washington, DC: The Community College Press.

Rugg, D.S. (1981). The Midwest as a hearth area in American academic geography. In B.W. Blouet (Ed.), *The origins of academic geography in the United States* (pp. 175–191). Hamden, CT: Archon Books.

Sabin, P. (1997). The two "Wests" of twentieth-century United States history. *Pacific historical review*, 66 (3), 305–335.

Sauer, C.O. (1925). The morphology of landscape. In C.O. Sauer (Ed.), *University of California publications in geography, Vol. 2* (pp. 19–53). Berkeley, CA: University of California Press.

Sauer, C.O. (1952). *Agricultural origins and dispersals*. New York: American Geographical Society.

Schaefer, F.K. (1953). Exceptionalism in geography. *Annals of the Association of American Geographers*, 43, 226–249.

Schmieder, O. (1928). The Russian colony of the Guadalupe Valley. *University of California publications in geography*, 2 (14), 409–434.

Semple, E.C. (1910). The Anglo-Saxons of the Kentucky mountains: A study in anthropogeography. *Bulletin of the American Geographical Society*, 42, 1–31

Shelley, F.M., Bean, L.L., and Shaw, S.J. (2001). SWT Geography: From the Texas Hill Country to international recognition. *The Southwestern Geographer, 5*, 61–79.

Sherman, J.H. (1991). In memoriam: G. Donald Hudson, 1897–1989. *Annals of the Association of American Geographers*, 81 (3), 515–518.

Shortridge, P., et al. (2008). Walter Martin Johannes Kollmorgen, 1907–2008. *From Lindley Alumni newsletter of the Department of Geography, University of Kansas*, 5–9

Simpson, R.E., and Huke, R.E. (1988). One view of geography at Dartmouth College. In J.E. Harmon and T.J. Rickart, *Geography in New England: A special publication of the NewEngland/St. Lawrence Valley Geographical Society* (pp. 10–24).

Skyers, A.G. (2006) *Transformational leadership practices of New England community college presidents*. Unpublished doctoral dissertation, University of Bridgeport.

Sloan, R.H. (2009). *A quantitative study of the relationship between transformational and transactional leadership styles and strategic change within the State University of New York*. Unpublished doctoral dissertation, Capella University.

Smith, G.H. (1957). Roderick Peattie, geographer and romanticist, 1891–1955. *Annals of the Association of American Geographers*, 47 (1), 97–99.

Smith, G.H., and Taaffe, E.J. (n.d.). *A chronology of the department of geography*, The Ohio State University. Unpublished manuscript.

Smith, N. (1987). Academic war over the field of geography: The elimination of geography at Harvard, 1947–1951. *Annals of the Association of American Geographers, 77,* 155–172.

Smith, N. (2003). *American empire: Roosevelt's geographer and the prelude to globalization.* Berkeley, CA: University of California Press.

Spencer, J.E. (1972). George McCutcheon McBride, 1876–1971. *Geographical review,* 62 (3), 428–430.

Spencer, J. E. (1979). A geographer west of the Sierra Nevada. *Annals of the Association of American Geographers,* 69 (10), 46–52.

Stimson, R.J. 2009. Reginald George Golledge (December 6, 1937–May 29, 2009). *Geographical research,* 47 (3), 337–338.

Symanski, R. (2002). *Geography inside-out.* Syracuse, NY: Syracuse University Press.

Taaffe, E.J. (1979). In the Chicago area. *Annals of the Association of American Geographers,* 69 (1), 133–138.

Taaffe, E.J. (1990). Some thoughts on the development of urban geography in the United States during the 1950s and 1960s. *Urban geography,* 11 (4), 422–431.

Taylor-Sawyer, S. (2004). *Community college chief executive officers and the Transformational leadership style.* Unpublished doctoral dissertation, New Mexico State University.

Thornthwaite, C.W. (1961). American geographers: A critical evaluation. *The professional geographer,* 13 (5), 10–13.

Thrower, N.J. (2005). *In memoriam: Henry J. Bruman, Professor of Geography Emeritus.* Unpublished manuscript.

Trewartha, G.T. et al. (1978). *Geography at the University of Wisconsin-Madison.* Unpublished manuscript.

Trewartha, G.T. (1979). Geography at Wisconsin. *Annals of the Association of American Geographers,* 69 (1), 16–21.

Tuan, Y.F. (1974). *Topophilia: A study of environmental perception, attitudes, and values.* Englewood Cliffs, NJ: Prentice Hall.

University of South Carolina Department of Geography. (2003). Strategic Plan. Retrieved on December 7, 2004 from http://www.cla.sc.edu/geog/program/strategic.html.

Van Valkenburg, S. (1949). Wallace W. Atwood. *Geographical review,* 49 (4), 675-677.

Velikonja, J. (1994). Geography at the University of Washington. *Yearbook of the Association of the Pacific Coast Geographers,* 56, 1–23.

Verdigets, F.L. (2008). *Insights into leadership: The role, political skill, social skill, and self-monitoring in mediating the relationship between transformational leadership and job performance.* Unpublished doctoral dissertation, Auburn University.

Visher, S.S. (1953). Rollin D. Salisbury and Geography. *Annals of the Association of American Geographers,* 43 (1), 4–11.

Walker, H.J., and Richardson, M.E. (1994). In memoriam: Fred Bowerman Kniffen, 1900–1993. *Annals of the Association of American Geographers,* 84 (4), 732–743.

Wallach, B. (1998). In memoriam: James Parsons, 1915–1997. *Annals of the Association of American Geographers,* 88 (2), 316–328.

Wallach, B. (1999). A slightly varnished history of the Department of Geography at the University of Oklahoma. *The southwestern geographer, 3,* 56–69.

Ward, D. (1977). Andrew Hill Clark, 1911–1975. *Annals of the Association of American Geographers,* 67 (1), 145–148.

Warf, B., and Webster, D.S. (1980). *Quality rankings of Ph.D.–granting departments of geography.* University of California, Los Angeles. Unpublished manuscript.

Warman, H.J. (1977). Samuel Van Valkenburg, 1890–1976. *Annals of the Association of American Geographers,* 67 (2), 273–277.

Waterman, S. (2002). Scholar, manager, mentor, mensch: Saul B. Cohen. *Political geography* 21, 557–572.

West, R.C. (2004). In search of early Man in Mexico, 1946–1949. *Journal of the Southwest,* 46 (1), 117–119.

White, G.F. (174). Edward A. Ackerman, 1911–1973. *Annals of the Association of American Geographers*, 64 (2), 297–309.
Whittlesey, D.S. (1929). Sequent occupance. *Annals of the Association of American Geographers, 15*, 187–191.
Wilbanks, T.J. (2004). Geography and technology. In S.D. Brunn, S.L. Cutter, and J.W. Harrington (Eds.), *Geography and technology* (pp. 3–16). Dordrecht, The Netherlands: Kluwer Academic Publishers.
Williams, F.E. (1956). Lawrence Martin, 1880–1955. *Annals of the Association of American Geographers*, 46 (3), 357–364.
Williams, M. (1983). The apple of my eye: Carl Sauer and historical geography. *Journal of historical geography*, 9 (1), 1–28.
Wishart, D. (2000). In memoriam: Leslie Hewes, 1906–1999. *Annals of the Association of American Geographers*, 90 (3), 573–578.
Wright, R., and Koch, N. (n.d.). *Geography in the Ivy League*. Unpublished manuscript.
Wrigley, G.M. (1951). Isaiah Bowman. *Geographical review*, 41 (1), 7–65.
Yacher, L. (1982). Erwin Josephus Raisz, 1893–1968. In G.J. Martin (Ed.), *Geographers: Biobibliographical Studies* (pp. 93–98). New York, Continuum.

Index

academic geographers: Abler, Ron, 126n8, 135, 145n14; Ackerman, Ed, 86, 87, 140; Adams, John, 45, 135, 145n14, 145n15; Angel, David, 145n9; Aschmann, Homer, 109n2; Atwood, Wallace, 36, 39n3, 51, 67–72, 84, 86, 90, 91n1, 145n17, 160, 166n2, 169, 179n5; Baker, O.E., 70, 179n5; Barrows, Harlan H., 22, 23, 24, 24n3, 41, 128, 139; Baugh, Ruth, 58n3, 111–112; Bennett, Charles, 150; Berry, Brian, 33n2, 91n5, 96, 102, 104n5, 104n6, 140–141, 142, 144n3, 144n6, 145n7, 147n34, 147n35, 147n36, 148n37, 148n38, 148n39; Beyers, William, 136; Bingham, Hiram, 14; Blanchard, Raoul, 84; Borchert, John, 44, 45, 46, 47n2, 128, 132, 160, 164; Bowden, Martyn, 133; Bowman, Isaiah, 14, 16, 24n2, 72n1, 73–76, 77–78, 79–80, 81n1, 88–89, 91n3, 91n4, 112, 113, 116n1, 117, 160, 166n1, 166n2, 169, 178n2; Bowman, Robert, 74, 106, 109n1, 109n2; Brand, Donald, 37; Brigham, Albert Perry, 16; Broek, Jan, 44, 46, 145n13, 160, 166n2, 168; Brown, Lawrence, 136; Brown, Ralph, 41, 43–44, 94, 99; Bruman, Henry, 115, 123, 170, 171; Bryan, Kirk, 84; Bunge, William, 102; Butzer, Karl, 26–33, 128, 129, 141, 144n3; Carlson, Fred, 49, 50; Carter, George, 74–75, 76–77, 78, 79–80, 81n2, 137, 146n28, 160, 166n2, 171; Chamberlain, James, 111, 115; Chamberlin, T.C., 14, 25, 35; Church, Phil, 100; Clark, Andrew, 31, 32, 33n4, 38, 74, 128, 132, 160, 173; Clark, Gordon, 148n42; Clark, William, 131, 145n8; Cohen, Saul, 84, 85, 133, 145n17, 164, 166n4, 167, 169; Colby, Charles, 22–23, 23, 24, 37, 41, 139, 179n5; Conzen, Michael, 141, 148n40; Cowles, Henry Chandler, 21, 24n1, 24n3, 41, 65n1, 118; Cressey, George, 61–64, 65n1, 125n1, 128, 129, 163, 164, 170, 179n9; Dacey, Michael, 96, 102; Darby, H. Clifford, 147n35; Davis, Darrell H., 36, 41, 42, 44, 46, 128, 160, 168; Davis, William Morris, 14, 16, 17, 18, 67, 83, 90, 119; Deasy, George, 56, 134; De Blij, Harm, 150, 156n2; de Laubenfels, David, 146n21; Deskins, Donald, 138; Denevan, William, 33n4, 145n10, 168, 176; Dicken, Samuel, 41, 43, 44, 121; Dodge, Stanley, 37, 119; Dow, Maynard W., x, 63, 185, 191; Downs, Roger, 134; Dozier, Jeffrey, 151–152; Durand, Loyal, 29, 32; Earle, Frances, 100, 103n2, 104n4; Ekblaw, Elmer, 70, 179n5; Entrikin, Nicholas, 131; Espenshade, Edward, 94, 95, 96, 147n33; Estes, Jack, 150, 152; Finch, Vernor, 26, 27, 27–25, 32, 39n5, 58n1, 128, 159, 179n5; Garrison, William, 95, 97n3, 100, 101–103, 104n5, 151; Getis, Art, 102; Gilbert, Grove Karl, 17; Ginsburg, Norton, 141, 147n36, 148n37, 148n38, 148n39, 148n40; Glacken, Clarence, 144n6; Golledge, Reginald, ix, x, 108, 110n3, 136, 150, 151, 153, 155, 156n3, 175, 180n12; Goodchild, Michael, 155; Goode, J. Paul, 22, 41; Gosling, L.A. Peter, 138; Gottmann, Jean, 74, 78; Gould, Peter, 134, 146n22, 146n23, 173; Gregory, Herbert, 14, 17; Griess,

Phyllis, 56, 58n3, 134; Hagerstrand, Torsten, 96, 102; Hall, Robert, 37, 100; Hammond, Edwin, 31, 33n1, 34n8, 109n2, 126n4, 146n21; Hanson, Susan, 145n16; Harris, Chauncy, 21, 22, 141, 142, 147n36, 148n41; Hartshorne, Richard, 29, 30, 32, 33n2, 41, 41–43, 46, 47n1, 53n2, 58n3, 65, 66n4, 86, 94, 106–108, 128, 147n35, 160, 170, 172; Harvey, David, 76, 134, 137–138, 146n27, 173, 178n3; Hobbs, William Herbert, 35, 36, 39n1, 39n4, 118, 129; Hoffman, George, 38; Holloway, Ruliff, 117; Hooson, David, 144n6; Hudson, G. Donald, 93–95, 100–103, 104n4, 128, 129, 135, 164, 174; Huntington, Charles, 49, 52, 53; Huntington, Ellsworth, 14, 41, 113; Illick, J. Rowland, 61, 85; James, Preston, 36, 37, 38, 39n5, 40n6, 58n3, 63–64, 65, 66n2, 66n4, 66n5, 70, 119, 125, 146n19, 160, 169; Jefferson, Mark, 23; Jenks, George, 34n7, 146n24; Jenson, Robert, 134; Johnson, D.W., 67; Jones, Clarence, 70, 71, 95–96, 96, 97n1, 97n2, 159, 179n5; Jones, Wellington, 37, 41, 119; Kates, Robert, 133, 145n16, 173; Katz, Cindi, 146n27; Kemp, Harold, 85–86, 88, 90; Kenzer, Martin, 185; Kesseli, John, 119, 122, 126n4, 126n5, 144n4; King, Leslie, 134, 136; Kish, George, 38, 139, 147n32, 178n1; Kniffen, Fred, 24n1, 35, 39n2, 42, 100, 120–121, 121; Knowles, Anne, 157n5; Knox, James, 145n10; Koelsch, William, 83, 133, 145n7, 145n16; Kohn, Clyde, 38, 108; Lattimore, Owen, 74–81, 137–138, 146n28, 163; Leighly, John, 36, 119, 120, 126n5, 129, 131, 144n4, 144n5, 160, 171, 175; Ley, David, 134; Lewis, Peirce, 38, 56, 134, 146n23; Lindberg, James, 108, 109; Lukermann, Fred, 43, 45, 132, 145n14; Marble, Duane, 96, 102, 147n33, 147n35; Marbut, Curtis, 70, 179n5; Marcus, Melvin, 139, 147n31; Martin, Geoffrey, 88, 125, 143n1, 185; Martin, Howard, 100, 103n1, 103n2, 104n4, 160; Martin, Lawrence, 25, 40n6; Mather, E. Cotton, 45, 45–46, 47n2, 128; Mather, Kirtley, 84; Mayer, Harold, 23, 24n4, 148n37; McBride, George, 113–114, 115, 125n1, 163, 164, 170, 179n9; McBryde, F. Webster, 50, 56; McCarty, Harold, 105–106, 108–109, 110n3, 160, 166n2, 170; McClellan, Myrta, 58n3, 111–113, 115, 159; McCune, Shannon, 50; McMurry, KC, 37, 38, 39, 119, 138, 160; Meigs, Peveril, 121; Meinig, Donald, 44, 99–100, 103n1, 134, 146n20, 170, 173; Mercer, John, 134; Mikesell, Marvin, 115, 140, 141, 142; Miller, E. Willard, 50, 52, 56–58, 59n4, 59n5, 128, 134, 164, 167, 174; Morrill, Richard, 95, 96, 101, 102, 135; Murphy, Alexander, 143, 190; Murphy, Raymond, 55, 56, 58n1, 71; Nash, Peter, 85; Nellis, Duane, 191; Nelson, Howard, 85, 145n8; Newman, James, 134, 146n18; Nietschmann, Bernard, 144n6, 147n31; Nystuen, John, 102, 138; Olsson, Gunnar, 139, 147n31; Olmstead, Clarence, 31, 38; Panzer, Wolfgang, 119; Parsons, James, 33n4, 122, 125, 130–131, 132, 144n6, 147n35, 150, 160, 171, 175, 176, 179n7; Pattison, William, 141, 147n34; Peattie, Roderick, 30, 49–51, 53n2, 53n3, 70; Pelzer, Karl, 74; Penck, Albrecht, 119, 120; Pfeifer, Gottfried, 119; Philbrick, Allen, 23; Pitts, Forrest, 38; Platt, Robert, 23, 24, 41, 128, 139, 140; Porter, Philip, 145n15; Pred, Allan, 144n6; Prunty, Merle, 70; Raisz, Erwin, 34n6, 146n24; Randall, John, 50, 55; Raup, Hallock, 112, 119, 120; Renner, George, 100; Robinson, Arthur, 30–31, 32, 50, 52, 53n2, 128, 146n24, 172; Rodgers, Allan, 56, 57, 134, 146n23; Rostlund, Erhard, 17; Rushton, Gerard, 110n3; Russell,

R.J., 120; Salisbury, Rollin, 14, 17, 18, 21, 23, 24n1, 25, 35, 65n1, 67, 111, 118, 139, 160, 164, 167, 174, 175, 179n7; Sauer, Carl, 15, 16, 18, 24n1, 27, 30, 35, 36, 37, 38, 39n5, 41, 42, 74, 78, 87, 99, 103n1, 112, 114, 117–125, 125n1, 126n8, 128, 129, 129–130, 131, 132, 160, 163, 164, 167, 168, 170, 171, 174, 175, 176, 179n7, 179n9; Sauer, Jonathan, 176; Schaefer, Fred K., 105, 106–109, 170; Schmieder, Oskar, 119, 121, 125n3; Scott, Allen, 104n5, 131, 145n7, 145n9; Seeman, Albert, 100; Semple, Ellen Churchill, 16, 22, 36, 41, 58n3, 70, 112, 179n5; Shantz, Homer, 70, 179n5; Sherman, John, 34n7, 100, 103n3, 136; Simkins, Paul, 56, 134; Simonett, David, ix, 150–156, 160, 164, 166n5, 167, 174, 175; Smith, Guy-Harold, 31, 49, 50, 51, 52, 53, 159; Smith, J. Russell, 14; Smith, Neil, 88, 146n27; Smith, Terry, 153–154; Sopher, David, 146n20; Spencer, Joseph, 114–115, 131, 170, 171; Stoddart, David, 144n6; Stone, Kirk, 31, 37, 38; Taaffe, Ned, 52, 53, 95–96, 136–137, 146n24, 164, 167, 174; Tarr, Ralph, 25; Taylor, T. Griffith, 22, 23, 31; Thomas, Morgan, 135; Thornthwaite, Clarence, 46, 121, 147n35; Thrower, Norman, 34n7, 146n24, 150; Tobin, Graham, 146n26; Tobler, Waldo, 102, 104n6, 138–139, 147n31, 147n35, 151, 153, 157n6, 157n8, 169; Tower, Walter S., 18; Trewartha,Glenn, 26, 27, 30, 34n5, 39n5, 94, 128, 168; Tuan, Yi-Fu, 126n5, 129, 132, 144n3, 144n4, 145n11, 145n14, 147n35; Ullman, Edward, 22, 86, 90, 101, 103, 103n1, 104n5; Van Cleef 2n2f, Eugene, 49, 50, 51; Van Valkenburg, Sam, 70, 72, 179n5; Walker, Richard, 144n6; Ward, David, 33n2, 33n4, 34n5, 132, 145n11, 145n12, 179n10; Ward, Robert DeC., 36, 84; Weaver, John, 44, 132, 145n13; Wernstedt, Frederick, 56, 134; Whitaker, J. Russell, 29, 32, 43; Whitbeck, Ray, 25, 26, 27, 160; White, Gilbert, 66n4, 129, 139–140, 142, 143, 144n3, 148n37, 160, 164, 175; Whittlesey, Derwent, 42, 46, 47n1, 84–91, 112, 169, 179n5; Williams, Tony, 134; Wolman, Gordon, 137–138, 146n28, 147n29, 147n30, 163, 164; Wolpert, Julian, 134, 150; Wrigley, Gladys, 88; Yacher, Leon, 34n6; Zierer, Clifford, 114, 159; Zelinsky, Wilbur, 123, 134, 146n23

American Association of University Professors (AAUP), 69

Bergman, Gustav, 108–109

Boas, George, 78

Burns, James McGregor, 2, 11n1

Committee of Ten, 14, 17

Conant James, 87, 89–90, 91n5, 178n2

Fortas, Abe, 77

geographical societies, conferences and publications: American Geographical Society, 73, 91n4, 112, 113, 117, 156n3, 178n2; American Society for Professional Geographers, 57, 65, 66n5; *Annals of the Association of American Geography*, 42, 78, 83, 85, 106, 145n7, 145n16; Association of American Geographers, 1, 25, 35, 50, 52, 57, 58n3, 65, 66n4, 70, 73, 79, 80, 96, 129, 133, 137, 149, 156n3, 176, 180n13, 185; California Geographical Society, 33n4; *Earth and State*, 88–89; *Economic Geography*, 58n3, 70, 132, 145n16; Gamma Theta Upsilon, The International Geographical Honor Society, 180n13; International Geographical Union, 50, 73, 96; *Journal of Geography*, 111; "The Morphology of Landscape", 18, 118; National Council for Geographic Education 19n3; National Council of Geography Teachers, 50, 52;

Pennsylvania Geographical Society, 59n5
geography programs and departments: Berkeley, 1, 6, 9, 11n2, 14, 18, 27, 32, 33n1, 117–125, 131, 143, 144n6, 155, 160, 167, 186; Chicago, 1, 7, 16, 18, 21–24, 27, 32, 122, 138, 139–143, 167, 169, 175, 187; Clark, 1, 16, 18, 67–72, 122, 128, 132–134, 143, 145n9, 167, 169, 179n4, 179n5, 188; Columbia, 1, 139, 143; Georgia, 190; Harvard, 1, 2, 6, 11n7, 14, 18, 83–91, 143, 160, 169, 176, 178n2, 179n4, 189; Idaho, 173; Illinois, 9; Iowa, 8, 11n7, 105–109, 137, 160, 170, 189; Johns Hopkins, 9, 11n7, 18, 72n1, 73–81, 137–138, 160, 169, 170, 178n2, 188; Kansas, 8; Kentucky, 190; LSU, 8, 190; Michigan, 1, 6, 18, 35–39, 122, 138–139, 143, 157n7, 160, 168–169, 178n1, 187; Minnesota, 2, 7, 9, 41–47, 132, 160, 187; Nevada-Reno, 176; North Carolina-Charlotte, 176; Northwestern, 6, 93–97, 139, 143, 167, 169, 190; Ohio State, 2, 7, 9, 32, 49–53, 135–137, 157n7, 168, 188; Oklahoma, 190; Penn State, 2, 55–58, 134, 167, 188; Pennsylvania, 14, 138, 143; San Diego State University, 176; Santa Barbara, 6, 11n2, 149–156, 167, 174, 186; Stanford, 138; Syracuse, 2, 6, 61–65, 134, 160, 169, 179n4, 188; Texas State University, 176; UCLA, 6, 111–116, 131, 155, 163, 169, 170, 186; Washington, 7, 99–103, 135, 167, 168, 189; Wisconsin, 1, 6, 25–33, 122, 128, 132, 160, 175, 187; Yale, 1, 14, 17, 138, 143, 179n4, 186
geography's missed opportunities, 180n13
Gilman, Daniel Coit, 14

informal leadership, 173

Kroeber, Alfred, 120, 121
Kuhn, Thomas, 191n1

leadership behaviors: enigmatic, 159, 170, 171; laissez faire, 3, 39, 160, 168, 169, 172; transactional, 4, 33, 65, 71, 89, 160, 171, 172; transformational, 3, 4, 5–6, 24, 46, 53, 64, 65, 159, 160, 165, 167, 171, 172, 174, 175, 177, 178
Leakey, Richard E., 180n13

McCarthy witch hunt, 62, 77–78, 99, 163
military experience and leadership behaviors, 179n8
Multifactor Leadership Questionnaire (MLQ), 174

National Center for Geographic Information and Analysis (NCGIA), 152, 154, 155
Nearing, Scott, 69

Office of Strategic Services (OSS), 29, 30, 33n4, 37, 86, 114, 115, 160, 172

Painter, Sydney, 78
paradigms in American geographic thought, 10, 14, 18, 118, 127, 135, 143n1, 149, 152, 155
Post, Jerrold, 72n2

Quantitative Revolution, 149, 170

Rice, Alexander Hamilton, 84, 88, 91n4, 91n5
Roosevelt, Franklin, 74, 91n5, 140

social consciousness and servant leadership, 164, 166n5, 172, 179n9
Shalala, Donna, 179n10

Veblen, Thorstein, 69
Versailles Peace Conference, 73

Wilson, Woodrow, 65, 73, 180n13
World War Two, 1, 6, 50, 56, 61, 62, 74, 121, 127, 140, 160, 163, 170

About the Author

Raised in various parts of Europe and North America, **Michael S. DeVivo** enlisted in the Marine Corps after attending high school in Connecticut. Following periods of service with the Marines and Navy, he was licensed as a U.S. Merchant Marine Officer and took command of a number of vessels for the mineral and oil industry. He subsequently engaged in university studies and entered academe. Dr. DeVivo, currently professor of geography at Grand Rapids Community College, hosts the *Conversation with a Geographer* oral history series, often exploring the role of various leaders and leadership behaviors with those interviewed. In recent years, he has given a number of talks on leadership, as well as the history of geography and geopolitics, in Africa, Asia, and the United States.

www.ingramcontent.com/pod-product-compliance
Ingram Content Group UK Ltd.
Pitfield, Milton Keynes, MK11 3LW, UK
UKHW040910190926
475560UK00008B/81

9 780739 199121